SPACE POLICY AND EXPLORATION

Space Policy and Exploration

William N. Callmers
Editor

Nova Science Publishers, Inc.
New York

For permission to use material from this book please contact us:
Telephone 631-231-7269; Fax 631-231-8175
Web Site: http://www.novapublishers.com

LIBRARY OF CONGRESS CATALOGING-IN-PUBLICATION DATA

Space policy and exploration / William N. Callmers, editor.
p. cm.
ISBN 978-1-60456-448-8 (softcover)
1. Astronautics--Government policy--United States. 2. Outer Space--Exploration--Government policy--United States. I. Callmers, William N.
TL789.8.U5S5873 2008
629.40973--dc22 2008014851

Published by Nova Science Publishers, Inc. ◘ New York

CONTENTS

PREFACE

This new book presents a wide spectrum of in-depth analyses detailing the US space program including policy, the space stations, the shuttles, and space exploration.

Chapter 1 - The "space age" began on October 4, 1957, when the Soviet Union (USSR) launched Sputnik, the world's first artificial satellite. Some U.S. policymakers, concerned about the USSR's ability to launch a satellite, thought Sputnik might be an indication that the United States was trailing behind the USSR in science and technology. The Cold War also led some U.S. policymakers to perceive the Sputnik launch as a possible precursor to nuclear attack. In response to this "Sputnik moment," the U.S. government undertook several policy actions, including the establishment of the National Aeronautics and Space Administration (NASA) and the Defense Advanced Research Projects Agency (DARPA), enhancement of research funding, and reformation of science, technology, engineering and mathematics (STEM) education policy.

Following the "Sputnik moment," a set of fundamental factors gave "importance, urgency, and inevitability to the advancement of space technology," according to an Eisenhower presidential committee. These four factors include the compelling need to explore and discover; national defense; prestige and confidence in the U.S. scientific, technological, industrial, and military systems; and scientific observation and experimentation to add to our knowledge and understanding of the Earth, solar system, and universe. They are still part of current policy discussions and influence the nation's civilian space policy priorities — both in terms of what actions NASA is authorized to undertake and the appropriations each activity within NASA receives. NASA has active programs that address all four factors, but many believe that it is being asked to accomplish too much for the available resources.

Further, the United States faces a far different world today. No Sputnik moment, Cold War, or space race exists to help policymakers clarify the goals of the nation's civilian space program. The Hubble telescope, *Challenger* and *Columbia* space shuttle disasters, and Mars exploration rovers frame the experience of current generations, in contrast to the Sputnik launch and the U.S. Moon landings that form the experience of older generations. As a result, some experts have called for new 21st century space policy objectives and priorities to replace those developed 50 years ago.

The authorization of NASA funding in the National Aeronautics and Space Act of 2005 (P.L. 109-55) extends through FY2008. Congress may decide to maintain or shift NASA's priorities during the next reauthorization. For example, if Congress believes that national prestige should be the highest priority, they may choose to emphasize NASA's human exploration activities, such as establishing a Moon base and landing a human on Mars. If they consider scientific knowledge the highest priority, unmanned missions and other science-related activities may be Congress' major goal for NASA. If international relations are a high priority, Congress might encourage other nations to become equal partners in NASA's activities. If spinoff effects, such as the creation of new jobs and markets and its effect on STEM education are Congress' priorities, then technological development, linking to the needs of business and industry, and education may become NASA's primary goals.

Chapter 2 - Launching satellites into orbit, once the exclusive domain of the U.S. and Soviet governments, today is an industry in which companies in the United States, Europe, China, Russia, Ukraine, Japan, and India compete. In the United States, the National Aeronautics and Space Administration (NASA) owns and launches its space shuttle. Private sector companies provide launch services for other NASA launches and most Department of Defense (DOD) launches. Commercial customers purchase launch services from the U.S. companies or their competitors. Since the early 1980s, Congress and successive Administrations have taken actions, including passing several laws, to facilitate the U.S. commercial space launch services business. The Federal Aviation Administration (FAA) regulates the industry.

Forecasts in the 1990s suggesting significant increases in launch demand sparked plans to develop new launch v ehicles. NASA and DOD created government-industry partnerships to develop new reusable launch vehicles (RLVs) and "evolved" expendable launch vehicles (EELVs), respectively. (The space shuttle is the only RLV today. All other launch vehicles are "expendable" — they can only be used once). Several U.S. private sector companies began developing their own launch vehicles. Projections for launch services demand declined dramatically beginning in 1999, however. NASA's efforts to develop a new RLV to replace the shuttle faltered. DOD's new EELVs (Atlas V and Delta IV) began service, but, with reduced demand, the companies that build them (Lockheed Martin and Boeing) want more DOD funding to defray their costs. In 2005, the two companies announced plans to merge their EELV launch services for U.S. government customers. The joint venture, if approved by regulatory authorities, would be named the United Launch Alliance. Commercial launch services would not be affected.

Congress is debating the future of the space shuttle, which returned to flight in July 2005 after a two and one-half year hiatus following the 2003 *Columbia* tragedy. President Bush directed NASA to terminate the shuttle in 2010, but some want it to continue until a replacement is available. NASA also is embarking on a program to build two new "shuttle-derived" launch vehicles to implement the President's "Vision for Space Exploration" to return astronauts to the Moon and someday send them to Mars. The FY2007 request includes $837 million for the first of these new vehicles, the Crew Launch Vehicle.

In October 2004, Burt Rutan's SpaceShipOne suborbital spacecraft won the $10 million Ansari X-prize. Some believe this heralds an era of comparatively affordable space tourism. Congress passed a law in 2004 (P.L. 108-492) that creates a regulatory environment for space tourism. Concerns that China benefitted militarily from knowledge gained through commercial satellite launches in the 1990s led to changes in U.S. satellite export policy. The changes, especially returning control over such exports to the State Department from the Commerce Department, remain controversial because of what some claim is a negative impact on U.S. satellite manufacturing companies whose clients may choose European suppliers to avoid the U.S. export control regulations.

Chapter 3 - Congress continues to debate NASA's International Space Station (ISS), a permanently occupied facility in Earth orbit where astronauts live and conduct research. Congress appropriated approximately $35 billion for the program from FY1985-2005. The initial FY2006 ISS request was $2.180 billion: $1.857 billion for construction and operations and $324 million for research to be conducted by ISS crews. In a July budget amendment, NASA transferred $168 million for ISS Crew/Cargo Services to another part of the NASA budget and reduced the ISS request commensurately. The final version of the FY2006 appropriations bill that includes NASA (H.R. 2862) cuts $80 million from the originally submitted budget, and NASA now plans to spend $306 million, instead of $324 million, on ISS research in FY2006.

The space station is being assembled in Earth orbit. ISS segments, crews, and cargo are taken into orbit by Russian or U.S. spacecraft. ISS has been permanently occupied by successive "Expedition" crews rotating on 4-6 month shifts since November 2000. "Expedition 12" is now aboard. Cost growth and schedule delays have characterized the program since its inception. The grounding of the space shuttle fleet after the 2003 *Columbia* tragedy and the July 2005 *Discovery* "Return to Flight" mission is further affecting schedule, operations, and cost. Most of the remaining ISS segments are designed to be launched by the shuttle and construction therefore is suspended. President Bush's January 2004 "Vision for Space Exploration" also is affecting the ISS program. He directed that the shuttle program be terminated in 2010, and changed the focus of ISS-based research to only that which supports his "Moon/Mars" Vision instead of the broadlybased program that was planned.

Canada, Japan, and several European countries became partners with NASA in building the space station in 1988; Russia joined in 1993. Except for money paid to Russia, there is no exchange of funds among the partners. Europe, Canada, and Japan collectively expect to spend about $11 billion of their own money. A reliable figure for Russian expenditures is not available.

In 1993, when the current space station design was adopted, NASA said it would cost $17.4 billion for construction (not including launch or other costs). That estimate grew to $24.1-$26.4 billion, leading Congress to legislate spending caps on part of the program in 2000. The estimate then grew by almost another $5 billion, leading NASA (at White House direction) to cancel or indefinitely defer some hardware to stay within the cap. NASA exceeded the cap in FY2005, however.

Controversial since the program began in 1984, the space station has been repeatedly redesigned and rescheduled, often for costgrowth reasons. Congress has been concerned about the space station for that and other reasons. Twenty-two attempts to terminate the program in NASA funding bills were defeated, however (3 in the 106th Congress, 4 in the 105th Congress, 5 in the 104th, 5 in the 103rd, and 5 in the 102nd). Three other attempts in broader legislation in the 103rd Congress also failed.

Current congressional debate focuses on the impact of space shuttle-related delays, and the future of ISS in light of President Bush's new exploration initiative.

Chapter 4 - The International Space Station (ISS) program began in 1993, with Russia joining the United States, Europe, Japan, and Canada. Crews have occupied ISS on a 4-6 month rotating basis since November 2000.

The U.S. Space Shuttle, which first flew in April 1981, has been the major vehicle taking crews and cargo back and forth to ISS, but the shuttle system has encountered difficulties since the *Columbia* disaster in 2003. Russian Soyuz spacecraft are also used to take crews to and from ISS, and Russian Progress spacecraft deliver cargo, but cannot return anything to Earth, since they are not designed to survive reentry into the Earth's atmosphere. A Soyuz is always attached to the station as a lifeboat in case of an emergency.

President Bush, prompted in part by the *Columbia* tragedy, made a major space policy address on January 14, 2004, directing NASA to focus its activities on returning humans to the Moon and someday sending them to Mars. Included in this "Vision for Space Exploration" is a plan to retire the space shuttle in 2010. The President said the United States would fulfill its commitments to its space station partners, but the details of how to accomplish that without the shuttle were not announced.

The shuttle *Discovery* was launched on July 4, 2006, and returned safely to Earth on July 17. This was the first of 16 *post-Columbia* flights to the ISS that NASA plans to complete before retiring the shuttle in 2010. The shuttle *Atlantis* followed with a September 9 launch that resumed construction of the International Space Station before returning to earth September 21. *Discovery* returned to the ISS in December, returning to earth December 22 after a 13-day mission that included four space walks and continued work on the station.

On June 29, 2006, the House passed the FY2007 Science, Justice, Commerce and Related Agencies appropriations bill (H.R. 5672), which includes funding for NASA. The bill would have funded the Exploration Capabilities account, which includes the shuttle and the ISS, at $6.194 billion in FY2007. The Senate Appropriations Committee reported its version of H.R. 5672 (S.Rept. 109-280) on July 13, 2006, recommending $6.235 billion for Exploration Capabilities, but the Senate did not take action on the bill. NASA programs were funded for FY2007 in the Revised Continuing Appropriations Resolution, 2007 (H.J.Res. 20, P.L. 110-5). P.L. 110-5 funded Exploration Capabilities at $6.140 billion.

Chapter 5 - The 1958 National Aeronautics and Space Act specified that military space activities be conducted by the Department of Defense (DOD). DOD and the intelligence community manage a broad array of space activities, including launch vehicle development, communications satellites, navigation satellites (the Global Positioning System — GPS), early warning satellites to alert the United States to foreign missile launches, weather satellites, reconnaissance satellites, and developing capabilities to protect U.S. satellite systems and to deny the use of space to adversaries (called "space control" or "counterspace systems"). The 1990-1991 Persian Gulf War is dubbed by some as the first "space war" because support from space displayed great improvement over what was available during the previous major conflict, Vietnam. These systems continue to play significant roles in U.S. military operations. How to organize DOD and the intelligence community to work effectively on space programs has been an issue for many years.

Tracking the DOD space budget is extremely difficult since space is not identified as a separate line item in the DOD budget. Additionally, DOD sometimes releases only partial information (omitting funding for classified programs) or will suddenly release without explanation new figures for prior years that are quite different from what was previously reported. Figures provided to CRS show a total (classified and unclassified) DOD space budget of $19.4 billion for FY2003, $20 billion for FY2004, $19.8 billion for FY2005, and a request of $22.5 billion for FY2006. The actual FY2006 and proposed FY2007 budget figures are not yet available.

Two DOD space programs that have been particularly controversial are Space Radar (formerly Space-Based Radar — SBR) and TSAT (the transformational communications satellite program). The programs are controversial because their cost estimates are high, and Congress has been skeptical of those estimates and of DOD's ability to manage the programs successfully based on past program performance. Congress cut DOD's $226 million FY2006 request for Space Radar by $126 million and its $836 million FY2006 request for TSAT by $400 million. The FY2007 requests for those programs are $266 million for Space Radar and $867 million for TSAT.

Chapter 6 - On August 9, 2005, the space shuttle *Discovery* successfully completed the first of two "Return to Flight" (RTF) missions — STS-114. It was the first shuttle launch since the February 1, 2003, *Columbia* tragedy. NASA announced on July 27, 2005, the day after STS-114's launch, that a second RTF mission would be indefinitely postponed because of a problem that occurred during *Discovery*'s launch that is similar to what led to the loss of *Columbia.* The next launch is currently expected some time in 2006. This chapter discusses the *Columbia* tragedy, the *Discovery* mission, and issues for Congress regarding the future of the shuttle. For more information, see CRS Issue Brief IB93062, *Space Launce Vehicles: Government Activities, Commercial Competition, and Satellite Exports*, by Marcia S. Smith.

Chapter 7 - On January 14, 2004, President George W. Bush announced new goals for the National Aeronautics and Space Administration (NASA), directing the agency to focus on returning humans to the Moon by 2020, and eventually sending them to Mars and "worlds beyond." The President invited other countries to join. Most of the funding for this "Vision for Space Exploration" is to be redirected from other NASA activities, including terminating the space shuttle program in 2010, and ending U.S. participation in the International Space Station by 2016. NASA released an implementation plan for the Vision on September 19,

2005, and estimated the cost of returning humans to the Moon by 2018 (NASA's current goal) at $104 billion. An estimate for sending people to Mars was not provided. This chapter identifies issues Congress has been considering as it debates the President's Vision.

Chapter 8 - The Department of Defense's (DOD's) programs to develop new satellites to alert U.S. military commanders to foreign missile launches, and to support missile defense objectives, are controversial because of cost growth and schedule slippage. SBIRS-High, managed by the Air Force, would replace existing Defense Support Program "early warning" satellites. The Space Tracking and Surveillance System (STSS, formerly SBIRS-Low), managed by the Missile Defense Agency, would perform missile tracking and target discrimination for missile defense objectives. The SBIRS-High program has breached Nunn-McCurdy cost growth limits several times, resulting in several program restructurings. This is the final edition of this report.

Chapter 9 - The National Aeronautics and Space Administration (NASA) estimates that without a servicing mission to replace key components, the Hubble Space Telescope will cease scientific operations in 2008 instead of 2010. In January 2004, then-NASA Administrator Sean O'Keefe announced that the space shuttle would no longer be used to service Hubble. He indicated that this decision was based primarily on safety concerns in the wake of the space shuttle *Columbia* accident in 2003. Many critics, however, saw it as the result of the new Vision for Space Exploration, announced by President Bush in January 2004, which focuses NASA's priorities on human and robotic exploration of the solar system. Hubble supporters hope to reverse the decision and proceed with a shuttle servicing mission. Dr. Michael Griffin, who became NASA Administrator in April 2005, has stated that he will reassess whether to use the shuttle to service Hubble after there have been two successful post-*Columbia* shuttle flights. Problems during the first flight in July 2005 led to postponement of the second, which is now planned for July 2006.

In: Space Policy and Exploration
Editor: W. N. Callmers, pp. 1-38

ISBN: 978-1-60456-448-8

Chapter 1

U.S. CIVILIAN SPACE POLICY PRIORITIES: REFLECTIONS 50 YEARS AFTER SPUTNIK*

Deborah D. Stine

ABSTRACT

The "space age" began on October 4, 1957, when the Soviet Union (USSR) launched Sputnik, the world's first artificial satellite. Some U.S. policymakers, concerned about the USSR's ability to launch a satellite, thought Sputnik might be an indication that the United States was trailing behind the USSR in science and technology. The Cold War also led some U.S. policymakers to perceive the Sputnik launch as a possible precursor to nuclear attack. In response to this "Sputnik moment," the U.S. government undertook several policy actions, including the establishment of the National Aeronautics and Space Administration (NASA) and the Defense Advanced Research Projects Agency (DARPA), enhancement of research funding, and reformation of science, technology, engineering and mathematics (STEM) education policy.

Following the "Sputnik moment," a set of fundamental factors gave "importance, urgency, and inevitability to the advancement of space technology," according to an Eisenhower presidential committee. These four factors include the compelling need to explore and discover; national defense; prestige and confidence in the U.S. scientific, technological, industrial, and military systems; and scientific observation and experimentation to add to our knowledge and understanding of the Earth,

* Excerpted from CRS Report RL34263, dated December 3, 2007.

solar system, and universe. They are still part of current policy discussions and influence the nation's civilian space policy priorities — both in terms of what actions NASA is authorized to undertake and the appropriations each activity within NASA receives. NASA has active programs that address all four factors, but many believe that it is being asked to accomplish too much for the available resources.

Further, the United States faces a far different world today. No Sputnik moment, Cold War, or space race exists to help policymakers clarify the goals of the nation's civilian space program. The Hubble telescope, *Challenger* and *Columbia* space shuttle disasters, and Mars exploration rovers frame the experience of current generations, in contrast to the Sputnik launch and the U.S. Moon landings that form the experience of older generations. As a result, some experts have called for new 21st century space policy objectives and priorities to replace those developed 50 years ago.

The authorization of NASA funding in the National Aeronautics and Space Act of 2005 (P.L. 109-55) extends through FY2008. Congress may decide to maintain or shift NASA's priorities during the next reauthorization. For example, if Congress believes that national prestige should be the highest priority, they may choose to emphasize NASA's human exploration activities, such as establishing a Moon base and landing a human on Mars. If they consider scientific knowledge the highest priority, unmanned missions and other science-related activities may be Congress' major goal for NASA. If international relations are a high priority, Congress might encourage other nations to become equal partners in NASA's activities. If spinoff effects, such as the creation of new jobs and markets and its effect on STEM education are Congress' priorities, then technological development, linking to the needs of business and industry, and education may become NASA's primary goals.

Current U.S. space policy is based on a set of fundamental factors which, according to an Eisenhower presidential committee, "give importance, urgency, and inevitability to the advancement of space technology."[1] These factors were developed fifty years ago as a direct result of the Soviet Union's (USSR) launch of the first artificial satellite, Sputnik. This launch began the "space age" and a "space race" between the United States and USSR.

The four factors are the compelling need to explore and discover; national defense; prestige and confidence in the U.S. scientific, technological, industrial, and military systems; and scientific observation and experimentation to add to our knowledge and understanding of the Earth, solar system, and universe.[2] They are still part of current policy discussions and influence the nation's civilian space policy priorities — both in terms of what actions NASA is authorized to undertake and the appropriations each activity within NASA receives.

NASA has active programs that address all four factors, but many believe that it is being asked to accomplish too much for the available resources. An understanding of how policy decisions made during the Sputnik era influence U.S. space policy today may be useful as Congress considers changing that policy. The response of Congress to the fundamental question, "Why go to space?," may influence NASA's programs, such as its earth-observing satellites, human exploration of the Moon and Mars, and robotic investigation of the solar system and wider universe as well as its policies on related activities, including spinoff technological development, science and mathematics education, international relations, and commercial space transportation.

This chapter describes Sputnik and its influence on today's U.S. civilian space policy, the actions other nations and commercial organizations are taking in space exploration, and why the nation invests in space exploration and the public's attitude toward it. The chapter concludes with a discussion of possible options for future U.S. civilian space policy priorities and the implication of those priorities.

SPUTNIK AND AMERICA'S "SPUTNIK MOMENT"

On October 4, 1957, the USSR launched Sputnik, the world's first artificial satellite. Sputnik (Russian for "traveling companion") was the size of a basketball and weighed 183 pounds (see figure 1). Sputnik's launch and orbit[3] still influences policy decisions 50 years later.

The USSR's ability to launch a satellite ahead of the United States led to a national concern that the United States was falling behind the USSR in its science and technology capabilities and thus might be vulnerable to a nuclear missile attack.[4] The resulting competition for scientific and technological superiority came to represent a competition between capitalism and communism.

Both the 85th Congress and President Eisenhower undertook an immediate set of policy actions in response to the launch of Sputnik. Congress established the Senate Special Committee on Space and Astronautics on February 6, 1958, and the House Select Committee on Science and Astronautics on March 5, 1958 — the first time since 1892 that both the House and Senate took action to create standing committees on an entirely new subject. Each committee was chaired by the Majority Leader. The Preparedness Investigating Subcommittee of the Senate Armed Services Committee was also active in analyzing the nation's satellite and missile programs.[5]

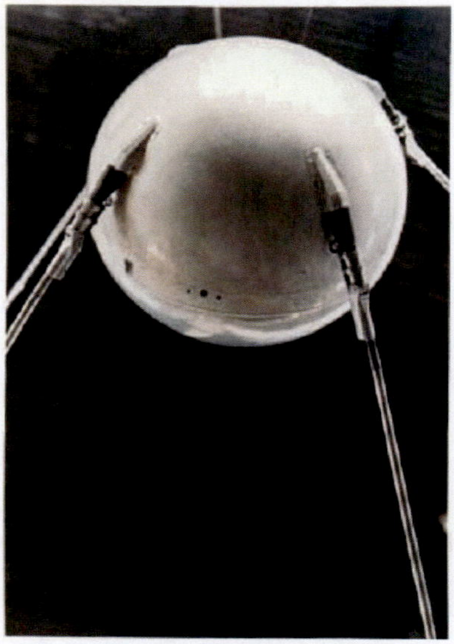

Source: NASA, at [http://history.nasa.gov/sputnik/gallerysput.html].

Figure 1. Sputnik.

Multiple congressional hearings were held in the three months following Sputnik, and President Eisenhower addressed the nation to assure the public that the United States was scientifically strong and able to compete in space. Within 10 months after Sputnik's launch, the Eisenhower Administration and Congress took actions that

- established the National Aeronautics and Space Administration (NASA) through the National Aeronautics and Space Act (P.L. 85-568),[6]
- established the Defense Advanced Research Projects Agency (DARPA) within the Department of Defense through DOD Directive 5105.15 and National Security - Military Installations and Facilities (P.L. 85-325),[7]
- increased its appropriation for the National Science Foundation to $134 million, nearly $100 million higher than the previous year,[8] and
- reformed elementary, secondary, and postsecondary science and mathematics education (including gifted education) and provided incentives for American students to pursue science, technology, engineering, and mathematics postsecondary degrees via fellowships and loans through the National Defense Education Act (P.L. 85-864).[9]

Figure 2 provides a timeline of the some of the major policy events in the year following the Sputnik launch.

When people today speak of a "Sputnik moment," they often refer to a rapid national response that quickly mobilizes major policy change as opposed to a response of inaction or incremental policy change. The term is also used to question inaction — as in whether or not the nation is prepared to respond to a challenge without an initiating Sputnik moment.

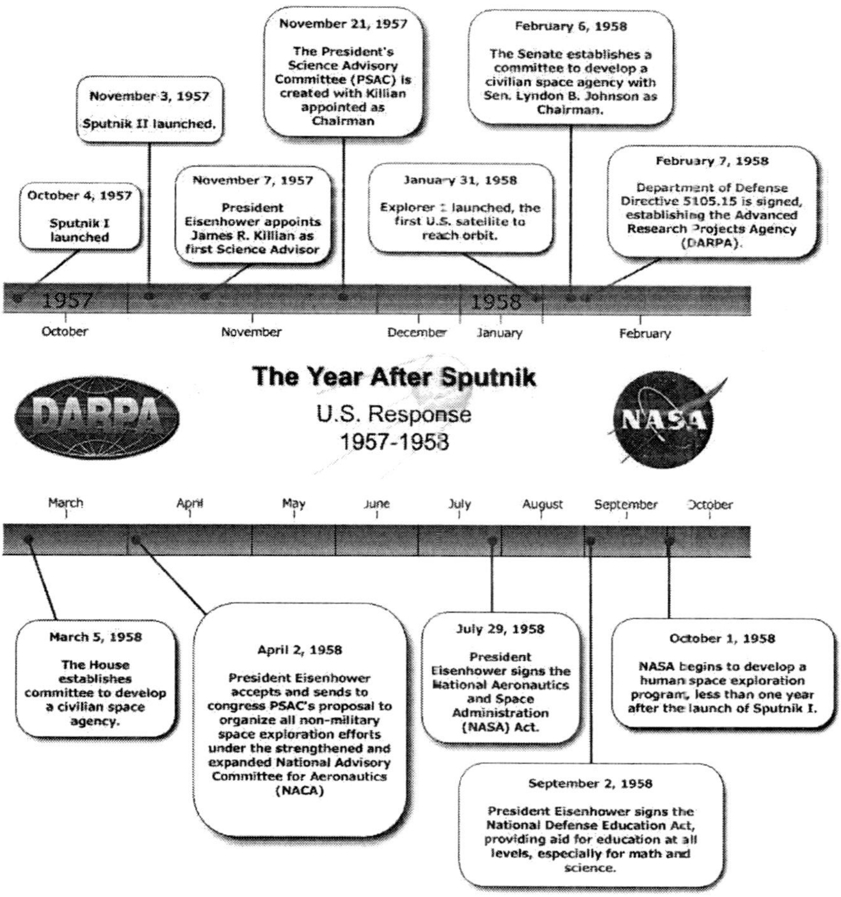

Source: Association of American Universities, at [http://www.aau.edu/education/Sputnik_Timeline_2007-09-20.pdf].

Notes: DARPA was also established by Congress in P.L. 85-325.

Figure 2. Timeline of Select Policy Events in the Year Following the Sputnik Launch.

WHY WAS SPUTNIK SO INFLUENTIAL?

The Sputnik launch captured the public's attention at a time of heightened U.S. tension regarding the threat posed by the USSR and communism. Societal focus on civil defense, including "duck and cover" drills and the establishment of some personal bomb shelters, predisposed the nation towards identifying the potential threat posed by the Sputnik launch.[10] In this climate, many Americans became concerned that if the USSR could launch a satellite into space, it could also launch a nuclear missile capable of reaching the United States.[11]

The Sputnik launch was immediately viewed as a challenge to U.S. scientific and technological prowess. The Soviet Union launched both Sputnik and Sputnik 2 before the United States was able to attempt a satellite launch.[12] Additionally, the Soviet launch was of a far heavier satellite than the U.S. had planned.[13] The net result of the Sputnik launch was called a "Pearl Harbor for American Science" — a sign that the United States was falling behind the USSR in science and technology.[14] The ensuing competition in scientific and technological skills came to represent a competition to determine the political superiority of capitalism versus communism.

The Senate Majority Leader at the time, future President Lyndon B. Johnson, illustrated the concern of many Americans in his own observations of the night sky: "Now, somehow, in some new way, the sky seemed almost alien. I also remember the profound shock of realizing that it might be possible for another nation to achieve technological superiority over this great country of ours."[15]

WHY IS SPUTNIK IMPORTANT TO TODAY'S POLICIES?

The Sputnik launch prompted rapid development of new federal policies and programs. In particular, federal investment in NASA is still influenced by the Sputnik-era principles as illustrated in the Space Act, both in terms of what actions NASA is authorized to undertake and the extent to which each activity is funded.

In 2005, NASA was reauthorized for FY2007 and FY2008.[16] As Congress considers future reauthorization of NASA and the agency's 50th anniversary in 2008, the status of the nation's space policy, and the relative importance of the various objectives underlying this policy may become topics of debate.

The United States faces a far different world today than 50 years ago. No Sputnik moment, Cold War, or space race exists to help policymakers clarify the goals of the nation's civilian space program. The Hubble telescope, *Challenger* and *Columbia* space shuttle disasters, and Mars exploration rovers frame the experience of current generations, in contrast to the Sputnik launch and the U.S. Moon landings that form the experience of older generations.

WHAT ARE THE ACTIVITIES OF OTHER NATIONS AND THE COMMERCIAL SECTOR IN SPACE EXPLORATION?

The United States faces a possible new set of competitors or collaborators in civilian space exploration. China, India, Japan, Russia, and Europe are taking an active role in space exploration as are commercial companies.[17] Japan's Moon orbiter was launched on September 14, 2007, and arrived at the Moon on October 4, 2007, the 50th anniversary of Sputnik. China's Moon orbiter was launched on October 24, 2007 and entered lunar orbit on November 5, 2007. India and the United States also plan to launch vehicles into the Moon's orbit by the end of 2008. The European Space Agency (ESA) is studying a Moon-lander mission, but its major focus is on Mars. The United Kingdom and Germany are independently proposing national Moon-orbiter projects, but their governments have not yet funded or approved these activities.[18]

China became the third country to send people into space, after the USSR[19] and the United States, completing two human spaceflight missions in 2003 and 2005.[20] Press reports indicate that China plans a manned Moon landing leading up to establishment of a Moon base after 2020,[21] though some have questioned these reports.[22] NASA Administrator Michael Griffin contends that China is likely to be the first country to land a human on the Moon since the last U.S. landing in 1972.[23] China has also indicated plans for a joint mission with Russia to send a satellite to orbit Mars in 2009.[24]

In the commercial sector, the X-Prize Foundation announced on September 13, 2007, the Google Lunar X Prize ($30 million) that invites private teams from around the world to build a robotic rover capable of landing on the Moon.[25] Virgin Galactic, currently based in California with a spaceport under construction in New Mexico, has plans for SpaceShipTwo, a six-passenger spaceliner with test flights scheduled for 2007 and suborbital passenger services to follow in 2008.[26] In Europe, EADS-Astrium is developing a four-person spacecraft to make suborbital trips with possibly the first commercial flight in 2012.[27] According to press reports, a number of venture capitalists are also planning to build spaceships or develop private space programs.[28]

If China is the first to land humans on the Moon and establish a Moon base in the 21st century or the European Space Agency is the first to land humans on

Mars, will policymakers and the public view these activities as a loss in United States status and leadership? If so, what are the policy implications? Would such activities become this century's "Sputnik moment" that would spur further investment in U.S. space exploration activities? If not, how might this affect U.S. space policy priorities?

WHAT IS THE NATION'S CURRENT CIVILIAN SPACE POLICY?

On August 31, 2006, President Bush announced a new U.S. National Space Policy.[29] The space policy defined the key objectives of defense and civilian space policy. The new space policy incorporated key elements of the Vision for Space Exploration ("Vision"), often referred to as the Moon/Mars program, previously announced on January 14, 2004.

In the Vision, President Bush announced new goals for NASA.[30] The President directed NASA to focus its efforts on returning humans to the Moon by 2020 and eventually sending them to Mars and "worlds beyond."[31] The President further directed NASA to fulfill commitments made to the 13 countries that are its partners in the International Space Station (ISS). In the 2005 NASA authorization act (P.L. 109-155), Congress directed NASA to establish a program to accomplish the goals outlined in the Vision, which are that the United States

- Implement a sustained and affordable human and robotic program to explore the solar system and beyond;
- Extend human presence across the solar system, starting with a human return to the Moon by the year 2020, in preparation for human exploration of Mars and other destinations;
- Develop the innovative technologies, knowledge, and infrastructures both to explore and to support decisions about the destinations for human exploration; and
- Promote international and commercial participation in exploration to further U.S. scientific, security, and economic interests.[32]

More specifically, the Vision includes plans, via a strategy based on "long-term affordability," to

- return the Space Shuttle safely to flight (which has been accomplished),
- complete the International Space Station (ISS) by 2010 but discontinue its use by 2017,

- phase out the Space Shuttle when the ISS is complete by 2010,
- send a robotic orbiter and lander to the Moon,
- send a human expedition to the Moon (sometime between 2015-2020),
- send a robotic mission to Mars in preparation for a future human expedition, and
- conduct robotic exploration across the solar system.[33]

NASA is developing a new spacecraft called Orion (formerly the Crew Exploration Vehicle) and a new launch vehicle for it called Ares I (formerly the Crew Launch Vehicle). An Earth-orbit capability is planned by 2014 (although NASA now considers early 2015 more likely) with the ability to take astronauts to and from the Moon following no later than 2020.

The Vision has broad implications for NASA, especially since almost all the funds to implement the initiative are expected to come from other NASA activities. Among the issues Congress is debating are the balance between NASA's exploration activities and its other programs, such as science and aeronautics research; the impact of the Vision on NASA's workforce needs; whether the space shuttle program might be ended in 2010; and if the United States might discontinue using the International Space Station.[34]

NASA states that its strategy is to "go as we can afford to pay," with the pace of the program set, in part, by the available funding.[35] Affording such a program is challenging, however, with a 2006 National Research Council report finding "NASA is being asked to accomplish too much with too little." The report recommended that "both the executive and the legislative branches of the federal government need to seriously examine the mismatch between the tasks assigned to NASA and the resources that the agency has been provided to accomplish them and should identify actions that will make the agency's portfolio of responsibilities sustainable."[36]

At a House Committee on Appropriations hearing, Dr. Michael Griffin, NASA's Administrator, acknowledged that funding is a challenge, stating

> Many people are amazed by the things that NASA accomplishes but don't realize that our budget is only 0.6 percent of the entire federal budget of the United States. As one of the most internationally recognized agencies in the government, many people assume that NASA's budget is much higher. In reality, we have to make tough choices in the allocation of scarce resources. We just cannot do everything that our many constituencies would like us to do. We need to set carefully considered priorities of time, energy and resources, and for this we're guided by the NASA Authorization Act of

2005, our annual appropriations, presidential policy, and the decadal surveys of the National Academy of Sciences. [37]

The House Committee on Appropriations also expressed concerns about these issues, stating NASA "has too many responsibilities and not enough resources to accomplish them all."[38] Similarly, the Senate Committee on Appropriations stated

> NASA's vision for space exploration maps out an aggressive role for the United States in manned space exploration. However, the potential costs are substantial and will likely be very difficult to maintain at the current estimated funding levels. The Committee is concerned that NASA will neglect areas that only tangentially benefit, or do not fit within, the exploration vision. The Committee believes that NASA must work diligently to balance existing programs and priorities with its plans for the future.[39]

WHY INVEST IN SPACE EXPLORATION?

The Appendix table compares The National Aeronautics and Space Act of 1958 as amended ("Space Act"),[40] the oldest and most recent Presidential commission reports (Killian[41] and Aldridge[42]), the U.S. National Space Policy[43] ("Space Policy"), and comments by members of Congress in the 110th Congress on the issue of why the United States might explore space. The analyses identify the following reasons why the United States might explore space:

- knowledge and understanding,
- discovery,
- economic growth — job creation and new markets,
- national prestige, and
- defense.

Some also include the following reasons:

- international relations, and
- education and workforce development.

Although there is broad agreement on the reasons for space exploration, there is a great deal of variation in the details. Among the chief differences in these documents are the degree to which

- discovery is the major reason for space exploration as opposed to meeting needs here on Earth;
- creation of jobs and new markets should be a major focus of NASA activities as opposed to a side effect;
- science and mathematics education and workforce development should be a goal of NASA in addition to other federal agencies; and
- relationships with other countries should be competitive or cooperative regarding space exploration.

According to an analysis conducted by the Space Foundation, the global space industry in 2005 generated $180 billion in revenues.[44] (See figure 3.) Comparing the Aldridge Commission themes, the Space Policy goals, and the Space Act objectives on the issue of the relationship of the space program to economic growth provides some insights. While the Aldridge committee has a much broader view of the industries related to space exploration, focusing on the potential role of space exploration in job generation and new market development, the Space Act and Space Policy focus on only one sector, the aeronautical and space vehicle industry.

The two Presidential commissions have two key differences. One is the first theme outlined in the Sputnik-era Killian Committee report: "the compelling urge of man to explore and discover." This is quite different from the recent Aldridge Commission report, which, although indicating exploration and discovery should be among NASA goals, states that "exploration and discovery will perhaps *not* be sufficient drivers to sustain what will be a long, and at times risky, journey." The implication is that, today, solely responding to the challenge of going to the Moon or Mars is not sufficient to energize public support for space exploration.

The second key difference is the focus of the Aldridge Commission on economic growth as a proposed space exploration theme. The Aldridge Commission identifies the ability of investments in civilian space programs to generate new jobs within current industries and spawn new markets. The contribution that federal space investments make to the nation's economy was not a key factor identified by the Killian Committee.

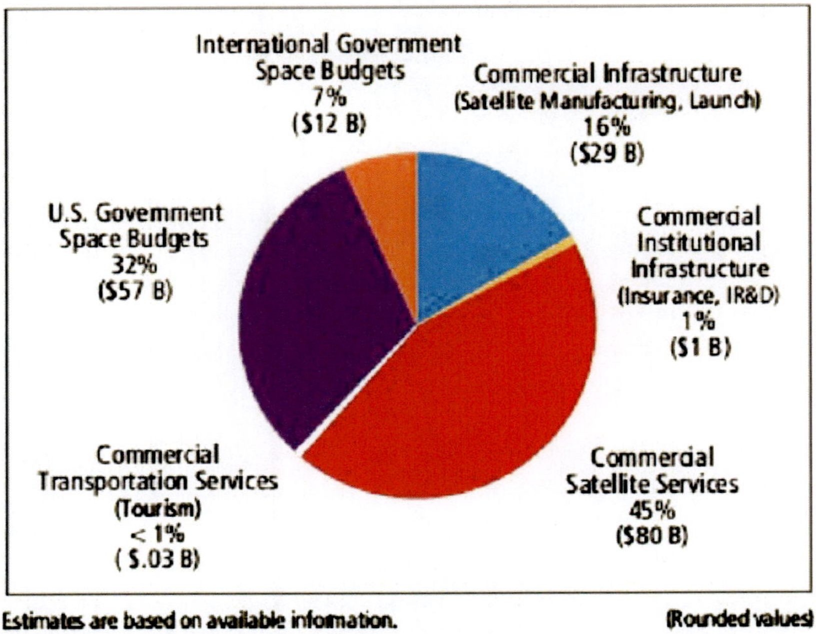

Source: Space Foundation, The Space Report: Guide to Global Space Activities, 2006, at [http://www.thespacereport.org/].

Figure 3. Global Space Economy Revenues, 2005.

As a result of its focus on economic growth as a key theme of space exploration, the Aldridge Commission recommended that "NASA's relationship to the private sector, its organizational structure, business culture, and management processes —all largely inherited from the Apollo era — must be decisively transformed to implement the new, multi-decadal space exploration vision." Two of its specific recommendations were that NASA recognize and implement a far larger private industry presence in space operations, with the specific goal of allowing private industry to assume the primary role of providing services to NASA, and that NASA's centers be reconfigured as Federally Funded Research and Development Centers (FFRDCs) to enable innovation, work effectively with the private sector, and stimulate economic development.[45]

FFRDCs are not-for-profit organizations which are financed on a sole-source basis, exclusively or substantially by an agency of the federal government, and not subject to Office of Personnel Management regulations. They operate as private non-profit corporations, although they are subject to certain personnel and budgetary controls imposed by Congress and/or their sponsoring agency. Each

FFRDC is administered by either an industrial firm, a university, or a nonprofit institution through a contract with the sponsoring federal agency. FFRDC personnel are not considered federal employees, but rather employees of the organization that manages and operates the center.[46]

Table 1. NASA Centers

Center	Mission Area	Location
Ames Research Center	New Technology Research	Moffett Field, CA
Dryden Flight Research Center	Flight Research	Edwards, CA
Glenn Research Center	Aeropropulsion and Communications Technologies	Cleveland, OH
Goddard Space Flight Center	Earth, the Solar System, and Universe Observations	Greenbelt, MD
Jet Propulsion Laboratory (FFRDC)	Robotic Exploration of the Solar System	Pasadena, CA
Johnson Space Center	Human Space Exploration	Clear Lake, TX, near Houston, TX
Kennedy Space Center	Prepare and Launch Missions Around the Earth and Beyond	Cape Canaveral, FL
Langley Research Center	Aviation and Space Research	Hampton, VA
Marshall Space Flight Center	Space Transportation and Propulsion Technologies	Huntsville, AL
Stennis Space Center	Rocket Propulsion Testing and Remote Sensing Technology	Hancock County, MS, near Slidell, LA

Note: FFRDC is a federally funded research and development center.
Source: NASA. [http://education.nasa.gov/about/nasacenters/index.html]

NASA has not fully adopted the Aldridge Commission recommendations. NASA has 10 centers (see table 1).[47] One, the Jet Propulsion Laboratory (JPL), is already an FFRDC and is managed by the California Institute of Technology. In 2005 testimony before Congress discussing the involvement of private industry in NASA activities, NASA Administrator Griffin identified a need for continued federal capabilities. In particular, on the issue of private entrepreneurs who believe that they can launch vehicles and put payloads, including human beings, into space at a fraction of what it costs NASA, Griffin stated, "while I am enlisting the entrepreneurial community to step forward and help meet those requirements, we cannot stop work on the, admittedly less efficient, government systems.... [T]hat just doesn't work.[48] Similarly, when asked about whether or not NASA Centers would be converted to FFRDCs, Administrator Griffin stated, "I do not fundamentally see any gain to be achieved by having NASA convert federal centers to FFRDCs."[49]

What Is the Public's Attitude toward Space Exploration?

Some editorialists question whether investing in space exploration is relevant today.[50] Others question if NASA has the right priorities.[51] Would the public care if the country's investment in space exploration ended? Does the public believe it would be better to invest in social needs here on Earth rather than space exploration? Does the public support the current prioritization of the nation's space exploration activities?

According to poll data, Americans do not rank space exploration as a high priority for federal government spending. For example, in an April 10, 2007 Harris poll, respondents were given a list of twelve federal government programs and asked to pick two which should be cut "if spending had to be cut." Space programs led the list (51%), followed by welfare programs (28%), defense spending (28%), and farm subsidies (24%).[52] Space exploration was also near the bottom of a University of Chicago National Opinion Research Center survey reported in January 2007 that asked Americans about how they would prioritize federal spending.[53]

On the other hand, Americans are interested in space exploration. According to a Zogby poll reported in May 2007, most Americans (66%) are interested in space exploration with about half of those personally interested in commercial space travel and tourism. About half (49%) gave NASA a positive job approval rating. Many of those surveyed (45%) expressed concerns that the shuttle fleet is too old, too expensive, and too frail to fly humans into space safely. NASA's great achievements, according to poll respondents, are the Apollo Moon missions (34%) and the Hubble Space Telescope (18%).[54]

NASA's Office of Strategic Communication funded several analyses of the public's attitude toward space exploration based on focus groups[55] and a survey,[56] the results of which were presented in June 2007.[57] According to an analysis conducted for NASA, the focus group participants were ambivalent about going to the Moon and Mars and wanted to know why these missions were important. Reasons such as leadership, legacy, and public inspiration were found to be less persuasive, especially for future Moon exploration, than NASA-influenced technologies. Most participants agreed that partnership with other countries would be beneficial, but doubted whether it can be achieved realistically.

In addition, one of the analysis conducted for NASA found that most survey respondents rated NASA-influenced technologies[58] as somewhat or extremely relevant to them. Over 52% of participants said such technologies were a "very strong" reason to go to space. In contrast, the public's response to a mission to send humans to the Moon by the year 2020 was less strong with 15% of respondents very excited and 31% somewhat excited. Results for a mission to send humans to the Mars were similar to those for the Moon.

The public opinion analysis conducted for NASA found that there are generational differences in regard to NASA's proposed activities. For example, NASA's base support came from those who encompass "The Apollo Generation" (45-64 year olds), the majority (79%) of whom support NASA's new space exploration mission, particularly the return to the Moon. By contrast, the majority (64%) of those between 18-24 years of age are uninterested or neutral about a human Moon mission. Those between 25 and 44 years of age are approximately evenly split between those who are interested/excited and those who are either uninterested or neutral. Those over 65 were more likely to be neutral or disinterested in a Moon mission, with those over 75 years of age the least interested of all age groups.[59]

What Are the Nation's Priorities for Civilian Space Exploration and Its Implications for Future Space Policy?

Current U.S. civilian space policy is based on a set of fundamental objectives in the Space Act, based on policy discussions that occurred following the launch of Sputnik over 50 years ago. Those objectives are still part of current policy discussions and influence the nation's civilian space policy priorities — both in terms of what actions NASA is authorized to undertake and the degree of appropriations each activity within NASA receives. NASA has active programs that address all its objectives, but many believe that it is being asked to accomplish too much for the available resources.

NASA was last reauthorized in 2005 for FY2007 and FY2008.[60] Thus, the reauthorization of NASA for FY2009 and beyond, along with NASA's upcoming 50th anniversary in Fall 2008, may provide an opportunity for Congress to rethink the nation's space policy. Some analysts have called for new 21st century space policy objectives and priorities to replace those developed 50 years ago.[61]

The goals of the nation's investment in space exploration may be a key factor in determining the focus of NASA's activities and the degree of funding appropriated for its programs. Congress and outside experts have concerns as to whether the United States can afford to implement President Bush's Vision for Space Exploration without adversely influencing NASA's other programs.[62]

Congress may need to make challenging decisions to determine how to reap the most benefit from the nation's civilian space program investment. These decisions might answer questions such as

- What are the priorities among the many reasons for U.S. space exploration? For example, what might be the priority ranking among the previously identified reasons as to why the United States might explore space — knowledge and understanding, discovery, economic growth, national prestige, defense, international relations, and education and workforce development?
- What implications would this prioritization have for NASA's current and future budgets and the balance among its programs? For example, what is the proper balance between human and robotic space activities?
- What influence might the timing of other countries' space exploration activities have on U.S. policy? For example, what would be the impact of the United States, China, or another country, or a commercial organization, establishing the first Moon base or landing on Mars?

New objectives and priorities might help determine NASA's goals. This, in turn, might potentially help Congress determine the most appropriate balance of funding available among NASA's programs during its authorization and appropriation process.

For example, if Congress believes that national prestige should be the highest priority, they may choose to emphasize NASA's human exploration activities, such as establishing a Moon base and landing a human on Mars. If they consider scientific knowledge the highest priority, Congress may emphasize unmanned missions and other science-related activities as NASA's major goal. If international relations are a high priority, Congress might encourage other nations to become equal partners in actions related to the International Space Station. If spinoff effects, including the creation of new jobs and markets and its catalytic effect on math and science education, are Congress' priorities, then they may focus NASA's activities on technological development and linking to the needs of business and industry, and expanding its role in science and mathematics education.

Appendix Table: Possible U.S. Civilian Space Policy Objectives: Comparison of Selected Extracts from Historical and Current Space Policy Documents

Possible U.S. Civilian Space Policy Objectives	Space Act Objectives (July 1958, *as amended*)	Eisenhower Administration "Killian Committee" Factors (March 1958)	G.W. Bush Administration "Aldridge Commission" Themes (2004)	G.W. Bush Administration "Space Policy" Goals (2006)	Congressional Comments on Civilian Space Policy Objective-Related Issues in the 110th Congress
Knowledge and Understanding	"(1) The expansion of human knowledge *of the Earth and* of phenomena in the atmosphere and space."	"Fourth, space technology affords new opportunities for scientific observation and experiment which will add to our knowledge and understanding of the earth, the solar system, and the universe."	"Exploring the Moon, Mars, and beyond is a great journey worthy of a great nation. The impulse to explore the unknown is a human imperative, and a notable part of what animates us as a people. This endeavor presents an opportunity to inspire a new generation of American explorers, scientist, entrepreneurs, and innovators."	"Increase the benefits of civil exploration, scientific discovery, and environmental activities."	"From 2010, for another 3 to 5 years, we will have no access in space. We are going to rely on the kindness of allies to go back. We cannot lose time or ground. Our national security and our national honor depend upon it. Also, this would have a tremendous impact on the state of science, which goes to major efforts in terms of better understanding our planet Earth, where we do suspect intelligent life, and also the impact of climate changes."[a] (Senator Barbara Mikulski)

Appendix Table (Continued).

Possible U.S. Civilian Space Policy Objectives	Space Act Objectives (July 1958, *as amended*)	Eisenhower Administration "Killian Committee" Factors (March 1958)	G.W. Bush Administration "Aldridge Commission" Themes (2004)	G.W. Bush Administration "Space Policy" Goals (2006)	Congressional Comments on Civilian Space Policy Objective-Related Issues in the 110th Congress
Discovery	"(2) The improvement of the usefulness, performance, speed, safety, and efficiency of aeronautical and space vehicles." "(3) The development and operation of vehicles capable of carrying instruments, equipment, supplies, and living organisms through space."	"The first of these factors is the compelling urge of man to explore and to discover, the thrust of curiosity that leads men to try to go where no one has gone before. Most of the surface of the Earth has now been explored and men now turn to the exploration of outer space as their next objective."	"Despite the spiritual, emotional, and intellectual appeal of a journey to space — exploration and discovery will perhaps *not* be sufficient drivers to sustain what will be a long, and at times risky, journey. We must also undertake this mission for pragmatic, but no less compelling reasons, which have everything to do with life here on Earth."	"Implement and sustain an innovative human and robotic exploration program with the objective of extending human presence across the solar system."	"...[H]ere we are on the 50th anniversary of Sputnik and it is another sputnik moment. When all of us in America were shocked that Russia had put up the first spaceflight, we were left to say: Why weren't we first? Today, 50 years later, we are looking at a 5-year gap from the end of the space shuttle before the crew-return vehicle will be on line to put American astronauts back in space. That is another Sputnik moment.

Possible U.S. Civilian Space Policy Objectives	Space Act Objectives (July 1958, *as amended*)	Eisenhower Administration "Killian Committee" Factors (March 1958)	G.W. Bush Administration "Aldridge Commission" Themes (2004)	G.W. Bush Administration "Space Policy" Goals (2006)	Congressional Comments on Civilian Space Policy Objective-Related Issues in the 110th Congress
					Are we going to rely on Russia after 2010 to put American astronauts in space? I hope not. I hope America never loses its commitment to be the first in technology, in knowing what can be done, in exploring issues we haven't even thought about because we know how much that exploration has already done for our country."[b] (Senator Kay Bailey Hutchinson)

Appendix Table (Continued).

Possible U.S. Civilian Space Policy Objectives	Space Act Objectives (July 1958, *as amended*)	Eisenhower Administration "Killian Committee" Factors (March 1958)	G.W. Bush Administration "Aldridge Commission" Themes (2004)	G.W. Bush Administration "Space Policy" Goals (2006)	Congressional Comments on Civilian Space Policy Objective-Related Issues in the 110th Congress
Economic Growth — Job Creation and New Markets	"(4) The establishment of long-range studies of the potential benefits to be gained from, the opportunities for, and the problems involved in the utilization of aeronautical and space activities for peaceful and scientific purposes."	Not discussed.	"Further space exploration will generate new jobs within current industries and will likely spawn entire new markets involving leading-edge manufacturing and flight support servicesAs one impressive labor leader testified to the Commission, 'every dollar spent on space is a dollar spent here on Earth.' This focus is good for jobs, good for the economy, and good for American families."	"Enable a dynamic, globally competitive domestic commercial space sector in order to promote innovation, strengthen U.S. leadership, and protect national, homeland, and economic security."	"NASA's space and Earth science basic research activities, along with microgravity research, are prime examples of research investments that cannot only advance our knowledge but benefit society [T]he investments also play a critical role in educating the next generation of scientists and engineers.

Possible U.S. Civilian Space Policy Objectives	Space Act Objectives (July 1958, as amended)	Eisenhower Administration "Killian Committee" Factors (March 1958)	G.W. Bush Administration "Aldridge Commission" Themes (2004)	G.W. Bush Administration "Space Policy" Goals (2006)	Congressional Comments on Civilian Space Policy Objective-Related Issues in the 110th Congress
					Aeronautics R and D, we have discussed that at some great length, is another area where investments we make benefit our economy, our quality of life, and our national security Human space flight and exploration is another area that offers benefits ranging from the very important intangible inspiration it provides to the public, to the advanced technologies, and research results that can come from those initiatives"[c] (Congressman Mark Udall)

Appendix Table (Continued).

Possible U.S. Civilian Space Policy Objectives	Space Act Objectives (July 1958, as amended)	Eisenhower Administration "Killian Committee" Factors (March 1958)	G.W. Bush Administration "Aldridge Commission" Themes (2004)	G.W. Bush Administration "Space Policy" Goals (2006)	Congressional Comments on Civilian Space Policy Objective-Related Issues in the 110th Congress
National Prestige	"(5) The preservation of the role of the United States as a leader in aeronautical and space science and technology and in the application thereof to the conduct of peaceful activities within and outside the atmosphere." "(9) The preservation of the United States' preeminent position in aeronautics and space through research and technology development related to associated manufacturing processes."	"Third, there is the factor of national prestige. To be strong and bold in space technology will enhance the prestige of the United States among the peoples of the world and create added confidence in our scientific, technological, industrial, and military strength."	"Exploring the Moon, Mars, and beyond is a great journey worthy of a great nation. The impulse to explore the unknown is a human imperative, and a notable part of what animates us as a people."	"Strengthen the nation's space leadership and ensure that space capabilities are available in time to further U.S. national security, homeland security, and foreign policy objectives."	"We are in a space race. While we are the current leader in space, there are many countries that want to take our place and are aggressively moving forward to do so...[O]ur Nation's space program benefits the lives of every American. The work that NASA does, from encouraging students into science and engineering careers, to innovative technology advances, improve our quality of life.

Possible U.S. Civilian Space Policy Objectives	Space Act Objectives (July 1958, as amended)	Eisenhower Administration "Killian Committee" Factors (March 1958)	G.W. Bush Administration "Aldridge Commission" Themes (2004)	G.W. Bush Administration "Space Policy" Goals (2006)	Congressional Comments on Civilian Space Policy Objective-Related Issues in the 110th Congress
					The forward and innovative thinking at NASA helps to ensure our Nation has the ability to compete, and lead, in the global economy. We are committed to keeping our leadership role in space. In order to do so, we must make the right investments in space at the right times. That time is now."[d] (Senator Richard Shelby)

Appendix Table (Continued).

Possible U.S. Civilian Space Policy Objectives	Space Act Objectives (July 1958, as amended)	Eisenhower Administration "Killian Committee" Factors (March 1958)	G.W. Bush Administration "Aldridge Commission" Themes (2004)	G.W. Bush Administration "Space Policy" Goals (2006)	Congressional Comments on Civilian Space Policy Objective-Related Issues in the 110th Congress
Defense	"(6) The making available to agencies directly concerned with national defense of discoveries that have military value or significance, and the furnishing by such agencies, to the civilian agency established to direct and control nonmilitary aeronautical and space activities, of information as to discoveries which have value or significance to that agency."	"Second, there is the defense objective for the development of space technology. We wish to be sure that space is not used to endanger our security. If space is to be used for military purposes, we must be prepared to use space to defend ourselves."	"Much of the United States' current military strength and economic security rests on our technological leadership. Our technological and industrial base must constantly be renewed. Therefore, the United States must continue to lead, especially in those industries that require, and therefore build, technology skills."	"Enable unhindered U.S. operations in and through space to defend our interests there." "Enable a robust science and technology base supporting national security, homeland security, and civil space activities."	"There is a strong, symbiotic relationship between space research and national security. For example, by using space-based navigation systems, we can guide a missile to within meters of its intended target. This not only allows our military to more effectively hit a target, it also saves civilian lives and limits collateral damage. The Chinese are gaining ground in technological areas. For example, China recently surpassed the U.S. as the world's largest exporter of information-technology products (and the U.S. has

Possible U.S. Civilian Space Policy Objectives	Space Act Objectives (July 1958, as amended)	Eisenhower Administration "Killian Committee" Factors (March 1958)	G.W. Bush Administration "Aldridge Commission" Themes (2004)	G.W. Bush Administration "Space Policy" Goals (2006)	Congressional Comments on Civilian Space Policy Objective-Related Issues in the 110th Congress
					become a net importer of those products). The Chinese are now turning their attention to space technology — and they are determined to use it as a means of strengthening their military. We cannot allow other countries to acquire new weapons technologies while America does not keep up."[c] (Senator Bill Nelson)

Appendix Table (Continued).

Possible U.S. Civilian Space Policy Objectives	Space Act Objectives (July 1958, as amended)	Eisenhower Administration "Killian Committee" Factors (March 1958)	G.W. Bush Administration "Aldridge Commission" Themes (2004)	G.W. Bush Administration "Space Policy" Goals (2006)	Congressional Comments on Civilian Space Policy Objective-Related Issues in the 110th Congress
International Relations	"(7) Cooperation by the United States with other nations and groups of nations in work done pursuant to this act and in the peaceful application of the results thereof."	Not discussed.	"Although the era of Sputnik has given way to an age of international cooperation in space, it remains a competitive frontier...Other nations, against whom we compete for jobs in the global economy, are also intent on exploring space. If not us, someone else will lead in the exploration, utilization, and ultimately, the commercialization of space, as we sit idly by."	"Encourage international cooperation with foreign nations and/or consortia on space activities that are of mutual benefit and that further the peaceful exploration and use of space, as well as to advance national security, homeland security, and foreign policy objectives."	"NASA has a successful history of international cooperation in science and involves non-U.S. partners on some two-thirds of its science missions, and also provides instruments, science support, and other in-kind contributions to non-U.S.-led space and Earth science missions. Successful cooperative missions can increase the scientific content of a mission and build mutually beneficial relationships.

Possible U.S. Civilian Space Policy Objectives	Space Act Objectives (July 1958, as amended)	Eisenhower Administration "Killian Committee" Factors (March 1958)	G.W. Bush Administration "Aldridge Commission" Themes (2004)	G.W. Bush Administration "Space Policy" Goals (2006)	Congressional Comments on Civilian Space Policy Objective-Related Issues in the 110th Congress
					At the same time, cooperation can lead to delays and added mission costs. Among the factors that have made international cooperative missions harder in recent years is ITAR" (International Traffic in Arms Regulations).[f] (House Committee on Science and Technology)

Appendix Table (Continued).

Possible U.S. Civilian Space Policy Objectives	Space Act Objectives (July 1958, as amended)	Eisenhower Administration "Killian Committee" Factors (March 1958)	G.W. Bush Administration "Aldridge Commission" Themes (2004)	G.W. Bush Administration "Space Policy" Goals (2006)	Congressional Comments on Civilian Space Policy Objective-Related Issues in the 110th Congress
Education and Workforce Development	"(8) The most effective utilization of the scientific and engineering resources of the United States, with close cooperation among all interested agencies of the United States in order to avoid unnecessary duplication of effort, facilities, and equipment."	Not discussed.	"Long-term competitiveness requires a skilled workforce. The space exploration vision can be a catalyst for a much-needed renaissance in math and science education in the United States."	Not discussed.	"Madam President, we are observing the 50th anniversary of the launch of Sputnik, the first artificial satellite that was launched by humans...we suddenly became shocked at the fact that we were falling behind in math, in science and technology, and that, lo and behold, with the symbolic value of the Soviet Union — at that point our mortal enemy in the Cold War — having achieved that first. Now, there is a lesson in what I have just discussed about our history in space that would teach us not to repeat that now. What is that lesson?

Possible U.S. Civilian Space Policy Objectives	Space Act Objectives (July 1958, as amended)	Eisenhower Administration "Killian Committee" Factors (March 1958)	G.W. Bush Administration "Aldridge Commission" Themes (2004)	G.W. Bush Administration "Space Policy" Goals (2006)	Congressional Comments on Civilian Space Policy Objective-Related Issues in the 110th Congress
					First of all, one of the great lessons of that era is the fact that we got excited about science and technology and mathematics and engineering and space flight. We produced a generation of exceptionally talented and educated young people who were told to go to their limit."[g] (Senator Bill Nelson)

Sources: "Space Act": P.L. 85-568, The National Aeronautics and Space Act, July 29, 1958. This analysis focuses on the objectives section. "Killian Committee": U.S. President (Dwight D. Eisenhower), President's Science Advisory Committee, *Introduction to Outer Space*, March 26, 1958. p. 2. Available at [http://www.hq.nasa.gov/office/pao/History/monograph10/doc6.pdf]. "Aldridge Commission": U.S. President (George W. Bush), President's Commission on Implementation of United States Space Exploration Policy, *A Journey to Inspire, Innovate, and Discovery*, June 2004, p. 11. Excerpts are from the section entitled "Why Go?". Available at [http://www.nasa.gov/pdf/60736main_M2M_report_small.pdf]. "Space Policy": U.S. President (G.W. Bush), U.S. National Space Policy, August 31, 2006, at [http://www.ostp.gov/html/US%20National%20Space%20Policy.pdf]. Excerpts are from section 3, "United States Space Policy Goals."

Appendix Table (Continued).

Notes: Excerpts are selected to reflect the general tone of text and are not necessarily the only language discussing these issues. The words in italics in the "Space Act" column show the changes made to the objectives since 1958.

[a] Senator Barbara Mikulski, "Departments of Commerce and Justice, and Science, and Related Agencies Appropriations Act, 2008," remarks in the Senate, Congressional Record, October 16, 2007, p. S12904.

[b] Senator Kay Bailey Hutchinson, "Departments of Commerce and Justice, and Science, and Related Agencies Appropriations Act, 2008," remarks in the Senate, Congressional Record, October 4, 2007, p. S12724.

[c] Congressman Mark Udall, U.S. Congress, House Committee on Science and Technology, Subcommittee on Space and Aeronautics, *NASA's Fiscal Year 2008 Budget Request*, opening statement, 110th Cong., 1st sess., March 15, 2007, H.Rept. 110-12 (Washington: GPO, 2007), p. 17.

[d] Senator Richard Shelby, "Departments of Commerce and Justice, and Science, and Related Agencies Appropriations Act, 2008," remarks in the Senate, Congressional Record, October 16, 2007, p. S12905.

[e]. Senator Bill Nelson, "Departments of Commerce and Justice, and Science, and Related Agencies Appropriations Act, 2008," remarks in the Senate, Congressional Record, October 4, 2007, p. S12726.

[f]. U.S. Congress, House Committee on Science and Technology, Subcommittee on Space and Aeronautics, *NASA's Space Science* Programs: Review of Fiscal Year 2008 Budget Request and Issues, 110th Cong., 1st sess., May 2, 2007, H.Rept. 110-24 (Washington: GPO, 2007), p. 6.

[g] Senator Bill Nelson, "Departments of Commerce and Justice, and Science, and Related Agencies Appropriations Act, 2008," remarks in the Senate, Congressional Record, October 4, 2007, p. S12726.

REFERENCES

[1] U.S. President (Dwight D. Eisenhower), President's Science Advisory Committee, *Introduction to Outer Space*, March 26, 1958. p. 1, at [http://www.hq.nasa.gov/office/pao/ History/monograph10/doc6.pdf].

[2] Ibid., pp. 1-2.

[3] Sputnik 1 orbited the Earth every 96 minutes until it fell from orbit on January 4, 1958, three months after its launch. Roger D. Launius, "Sputnik and the Origins of the Space Age," at [http://history.nasa.gov/sputnik/ sputorig.html].

[4] Dwight D. Eisenhower Presidential Center, "Sputnik and the Space Race," at [http://www.eisenhower.utexas.edu/dl/Sputnik/Sputnikdocuments.html].

[5] U.S. Congress, House Committee on Science and Technology, *Toward the Endless Frontier: History of the Committee on Science and Technology, 1959-79,* prepared for the Committee by Ken Hechler, committee print, 96th Cong., 2nd sess., H.Prt. 35-120, (Washington: GPO, 1980), pp. 1-28.

[6] P.L. 85-568, The National Aeronautics and Space Act ("Space Act"), July 29, 1958, at [http://www.nasa.gov/offices/ogc/about/space_act1.html].

[7] DARPA was originally called the Advanced Research Projects Agency (ARPA). It was established by DOD Directive 5105.15 on February 7, 1958, and by Congress in P. L. 85-325 on February 12, 1958. The name was changed from ARPA to DARPA by DoD Directive on March 23, 1972. DARPA was redesignated ARPA by President Bill Clinton in an Administration document on February 22, 1993. ARPA's name was changed back to DARPA by P.L. 104-106 on February 10, 1996. For more information about DARPA and its history, see DARPA, "Defense Advanced Research Project Agency: Technology Transition," January 1997 at [http://www.darpa.mil/body/pdf/transition.pdf].

[8] The appropriation for NSF continued to increase in future years. In 1968, it was almost $500 million. National Science Foundation, An Overview of the First 50 years, at [http://www.nsf.gov/about/history/overview-50.jsp].

[9] P.L. 85-864, National Defense Education Act (NDEA), September 2, 1958.

[10] Dwight D. Eisenhower Presidential Center, "Sputnik and the Space Race," at [http://www.eisenhower.utexas.edu/dl/Sputnik/Sputnikdocuments.html].

[11] Council on Foreign Relations, Chronology of National Missile Defense Programs, June 1, 2002, at [http://www.cfr.org/publication/10443/].

[12] The Project Vanguard booster tests on December 6, 1957 (rose 3 feet, caught fire) and February 5, 1958 (rose 4 miles, exploded) were unsuccessful. A new effort, Project Explorer led by Wernher von Braun,

was initiated. Explorer 1 was successful, after two aborted launches, on January 31, 1958. Roger D. Launius, "Sputnik and the Origins of the Space Age," at [http://history.nasa.gov/sputnik/sputorig.html].

[13] Sputnik 1 weighed 183 pounds. Sputnik 2 launched on November 3, 1957 weighed 1,120 pounds, carried a dog, and stayed in orbit for almost 200 days. The first satellite to be launched in the American Project Vanguard was planned to be 3.5 pounds. Roger D. Launius, "Sputnik and the Origins of the Space Age," at [http://history.nasa.gov/sputnik/sputorig.html].

[14] Davis, James C., *The Human Story: Our History, From the Stone Age to Today* (New York: Harper Collins, 2004). According to Davis, the statement was made by a Japanese newspaper shortly after the event. Others called it a "technological Pearl Harbor."

[15] Roger D. Launius, "Sputnik and the Origins of the Space Age," at [http://history.nasa.gov/sputnik/sputorig.html].

[16] NASA Authorization Act of 2005 (P.L. 109-155).

[17] Peter Spotts, "Many Contestants in Latest 'Space Race' to the Moon," *Christian Science Monitor*, October 1, 2007.

[18] Frank Morring, Jr., Michael A. Taverna and Neelam Mathew, "Nations Looking For a Piece of the Exploration Pie," *Aviation Week and Space Technology*, September 30, 2007, at [http://www.aviationweek.com/aw/generic/story_channel.jsp?channel= space and id=news/aw100107p2.xml and headline=Nations% 20Looking%20 For%20a%20 Piece% 20of%20the %20Exploration%20Pie].

[19] Since the fall of the USSR, Russia is the only former-USSR country to conduct space missions.

[20] See CRS Report RS21641, *China's Space Program: An Overview*, by Marcia S. Smith.

[21] *Agence France-Press*, "China Aims for Lunar Base after 2020," September 26, 2007, at [http://news.yahoo.com/s/afp/20070926/ wl_asia_afp/spacechinamoon_070926154203]. Paul Maidmen, "China Shoots for the Moon," *Forbes*, August 14, 2007, at [http://www.forbes.com/ opinions/2007/08/14/notn-china-aerospace-oped-cx_pm_0814no tn.html].

[22] Dwayne Day, "Exploding Moon Myths: Or Why There's No Race to Our Nearest Neighbor," The Space Review, November 12, 2007, at [http://www.thespacereview.com/ article/999/1].

[23] U.S. Congress, House Committee on Science and Technology, *NASA Fiscal Year 2008 Budget Request*, hearing, 110th Cong.,1st sess., March 15, 2007, H.Rept. 110-12, (Washington: GPO, 2007), p. 48. Michael D. Griffin, "The Role of Space Exploration in the Global Economy," speech, September 17,

2007, p. 41 at [http://www.nasa.gov/audience/formedia/speeches/ mg_ speech_collection_archive_1.html].

[24] "China and Russia Join Hands to Explore Mars," *People's Daily Online*, May 30, 2007, at [http://english.people.com.cn/200705/30/ eng20070530_ 379330.html].

[25] X-prize Foundation, "Google Sponsors Lunar X PRIZE to Create a Space Race for a New Generation," press release, September 13, 2007, at [http://www.xprize.org/lunar/ press-release/google-sponsors-lunar-x-prize-to-create-a-space-race-for-a-new-generation].

[26] Virgin Galactic, Overview, at [http://www.virgingalactic.com/]. NASA has signed a memorandum of understanding with Virgin Galactic to explore the potential for collaborations on the development of space suits, heat shields for spaceships, hybrid rocket motors, and hypersonic vehicles capable of traveling five or more times the speed of sound. See NASA, "NASA, Virgin Galactic to Explore Future Cooperation," press release, February 21, 2007, at [http://www.nasa.gov/home/hqnews/2007/feb/ HQ_07049_Virgin_ Galactic.html].

[27] EADS-Astrium, "Astrium Rockets into Space Tourism," press release, June 13, 2007, at [http://www.astrium.eads.net/press-center/press-releases/ astrium-rockets-into-space-touri sm].

[28] Jeremy Quittner (ed.), "I Need My Space," *Business Week*, Winter 2007, at [http://www.businessweek.com/magazine/content/07_09/b4023413.htm].

[29] U.S. President (G.W. Bush), *U.S. National Space Policy*, August 31, 2006, at [http://www.ostp.gov/html/US%20National%20Space%20Policy.pdf]. It replaced the previous space policy that had been in place for 10 years.

[30] U.S. President (G.W. Bush), *President Bush Announces New Vision for Space Exploration Program, Fact Sheet: A Renewed Spirit of Discovery*, January 14, 2004, at [http://www.whitehouse.gov/ news/releases/ 2004/01/ 20040114-1.html].

[31] Twelve U.S. astronauts walked on the Moon between 1969 and 1972. No humans have visited Mars.

[32] U.S. President (G. W. Bush), "A Renewed Spirit of Discovery," document, January 14, 2004, at [http://www.whitehouse.gov/space/ renewed_ spirit.html].

[33] NASA, *Vision for Space Exploration*, February 2004, at [http://www.nasa. gov/mission_pages/exploration/main/index.html].

[34] CRS Report RS22625, National Aeronautics and Space Administration: Overview, FY2008 Budget in Brief, and Key Issues for Congress, by Daniel Morgan and Carl E. Behrens; CRS Report RL33568, The International

Space Station and the Space Shuttle, by Carl E. Behrens; and CRS Report RS21720, Space Exploration: Issues Concerning the "Vision for Space Exploration," by Marcia S. Smith.

[35] Testimony of Michael D. Griffin, Administrator, National Aeronautics and Space Administration before the Senate Committee on Commerce, Science and Transportation Subcommittee on Space, Aeronautics and Related Sciences, Budget Hearing, U.S. Senate, February 28, 2007, at [http://commerce.senate.gov/public/_files/ Testimony_MichaelDGriffin_NASA_FY2008PostureStatementFINAL22707.pdf].

[36] National Research Council, Space Studies Board, *An Assessment of Balance in NASA's Science Program*, Washington, DC, 2006, p. 2 [http://www.nap.edu/ catalog.php?record_id=11644].

[37] Testimony of Michael D. Griffin, Administrator, National Aeronautics and Space Administration before the U.S. House of Representatives, Committee on Appropriations, Subcommittee on Commerce, Justice, Science, and Related Agencies, FY2008 Budget Hearing, March 13, 2007, at [http://www.cq.com/display.do?dockey=/cqonline/prod/data /docs/html/transcripts/congressional/110/congressionaltranscripts110-000002468892.ht ml@committees and metapub=CQ-CONGTRAN SCRIPTS].

[38] U.S. Congress, House Committee on Appropriations, report to accompany H.R. 3093, 110th Cong., 1st sess., July 25, 2007, H.Rept. 110-24, part 1 (Washington: GPO, 2007), p. 109.

[39] U.S. Congress, Senate Committee on Appropriations, report to accompany S. 1745, 110th Cong., 1st sess., June 29, 2007, S.Rept. 110-24, part 1 (Washington: GPO, 2007), p. 101.

[40] P.L. 85-568, The National Aeronautics and Space Act, July 29, 1958, at [http://www.nasa.gov/offices/ogc/about/space_act1.html]. Since 1958, the objectives have only had two modifications. The clause, "of the Earth and" was added to the first objective by the National Aeronautics and Space Administration Authorization Act, 1985, P.L. 98-361, § I 10(b), 98 Stat. 422, 426 (July 16, 1984). Objective (9) was added by the National Aeronautics and Space Administration Authorization Act, Fiscal Year 1989, P.L. 100-685, § 214, 102 Stat. 4083, 4093 (November 17, 1988). Objective (9) states the following: "The preservation of the United States' preeminent position in aeronautics and space through research and technology development related to associated manufacturing processes."

[41] During the Sputnik era, President Eisenhower's Science Advisory Committee, chaired by George Killian, ("Killian Commission") responded to the fundamental question of why the United States might undertake a national space program in its report *Introduction to Outer Space*. (U.S. President (Dwight D. Eisenhower), President's Science Advisory Committee, *Introduction to Outer Space*, March 26, 1958. p. 2, at [http://www.hq.nasa.gov/office/pao/History/monograph10/doc6.pdf]). The President's Science Advisory Committee is analogous to today's President's Council of Advisors on Science and Technology (PCAST).

[42] U.S. President (George W. Bush), President's Commission on Implementation of United States Space Exploration Policy, *A Journey to Inspire, Innovate, and Discovery*, June 2004. Available at [http://www.nasa.gov/pdf/60736main_M2M_report_small.pdf]. The commission report is named for its chair, Edward C. "Pete" Aldridge, Jr., and called the "Aldridge Commission" report.

[43] U.S. President (G.W. Bush), *U.S. National Space Policy*, August 31, 2006, at [http://www.ostp.gov/html/US%20National%20Space%20Policy.pdf].

[44] Space Foundation, The Space Report: Guide to Global Space Activities, 2006, at [http://www.thespacereport.org/]. For more on the space economy, see Michael D. Griffin, Administrator, National Aeronautics and Space Administration, *The Space Economy*, NASA 50th Anniversary Lecture Series, September 17, 2007, at [http://www.nasa.gov/audience/formedia/speeches/mg_speech_collection_archive_1.html].

[45] U.S. President (George W. Bush), President's Commission on Implementation of United States Space Exploration Policy, *A Journey to Inspire, Innovate, and Discovery*, June 2004, p. 7, at [http://www.nasa.gov/pdf/60736main_M2M_report_small.pdf].

[46] See CRS Report RS21542, Department of Homeland Security: Issues Concerning the Establishment of Federally Funded Research and Development Centers (FFRDCs), by Michael E. Davey.

[47] For more information, see [http://education.nasa.gov/about/ nasacenters/index.html].

[48] U.S. Congress, House Committee on Science, *Status of NASA's Programs*, H.Rept. 109-31, 109th Cong., 1st session, November 2, 2005, p. 53 (Washington, GPO, 2005).

[49] U.S. Congress, House Committee on Science, *The Future of NASA*, H.Rept. 109-19, 109th Cong., 1st session, June 28, 2005, p. 48 (Washington, GPO, 2005).

[50] See, for example, Anne Applebaum, "Mission to Nowhere," *Washington Post*, January 7, 2004, p. A21.

[51] See, for example, Gregg Easterbrook, "How NASA Screwed Up (And Four Ways to Fix It)," *Wired*, May 22, 2007, at [http://www.wired.com/science/space/magazine/ 15-06/ff_space_nasa]; *The Economist*, "Spacemen Are from Mars," September 27, 2007, at [http://www.economist.com/opinion/displaystory.cfm?story_id=9867224].

[52] Harris Interactive, "Closing the Budget Deficit: U.S. Adults Strongly Resist Raising Any Taxes Except "Sin Taxes" Or Cutting Major Programs," press release, April 10, 2007, at [http://www.harrisinteractive.com/ harris_poll/index.asp?PID=746]. The poll was of 2,223 adults surveyed online between March 6 and 14, 2007. This online survey is not based on a probability sample and therefore no theoretical sampling error can be calculated.

[53] University of Chicago, "Americans Want to Spend More on Education, Health," press release, January 10, 2007, at [http://www-news.uchicago.edu/releases/07/070110.gss.shtml]. The General Social Survey, supported by the National Science Foundation, has been conducted since 1973, and is based on face-to-face interviews of randomly selected people who represent a scientifically accurate cross section of Americans. For the 2006 survey, 2,992 people were interviewed and asked a wide variety of questions in addition to those related to spending priorities.

[54] Zogby, "Zogby Poll: Most Believe Humans Will Someday Colonize the Moon," press release, May 3, 2007, at [http://www.zogby.com/news/ ReadNews2.dbm?ID=1296]. The Zogby Interactive poll of 4,824 adults nationwide was conducted online from March 14-16, 2007, and carries a margin of error of ± 1.4 percentage points.

[55] [55] The focus groups were professionally moderated by Dr. Stephen Everett of the Everett Group, Inc., in consultation with ViaNovo. The six focus groups were located in San Diego, Kansas City, and Philadelphia.

[56] The professionally conducted telephone survey was of 1,001 U.S. adults in February 2007. The margin of error was ± 3.2%. The survey was conducted by Dr. Mary Lynne Dittmar of Dittmar Associates, in consultation with ViaNovo.

[57] Robert Hopkins, "Strategic Communications Framework Implementation Plan," powerpoint presentation, NASA, Office of Strategic Communications, June 26, 2007, at [http://www.spaceref.com/news/ viewsr.html?pid=24646].

[58] An example of a NASA-influenced technology (commonly called "spinoff") mentioned in the survey that had significant results is a smoke alarm. According to NASA, in the 1970s NASA needed a smoke and fire detector with adjustable sensitivity for Skylab, America's first space station. Honeywell developed the device for NASA and then made it available commercially so that consumers could avoid "nuisance" alarms while cooking. Other devices in the survey were advanced breast cancer imaging, heart defibrillators, weather satellites, remote-controlled robots, global positioning system, cordless tools, satellite radio, and DirecTV. See [http://www.sti.nasa.gov/tto/] for more details on NASA's spinoffs. See a list of NASA's top 20 spinoffs in the last five years at [http://www.ipp.nasa.gov/spinoff_top_20a.pdf].

[59] Ibid., p. 9. Robert Hopkins, "Strategic Communications Framework,", powerpoint presentation, NASA, Office of Communications Planning, February 2007, at [http://images.spaceref.com/news/2007/ feb 07.stratcomm.pdf]. M. L. Dittmar, *The Market Study for Space Exploration*, (Houston, TX: Dittmar Associates, Inc., 2004), pp. 26-29 (age data) and pp. 8-11 (Executive Summary). M. L. Dittmar, "Engaging the 18-25 Generation: Educational Outreach, Interactive Technologies, and Space". Paper #2006-7303 in Proceedings of AIAA Space 2006, September 19-21, (San Jose, California. Washington, D.C.: AIAA, 2006). Paper and presentation available at [http://www.dittmar-associates.com/ Paper_ Downloads.htm].

[60] P.L. 109-155, NASA Authorization Act of 2005, December 30, 2005.

[61] See, for example, International Academy of Astronautics, "The Next Steps in Exploring Deep Space," June 9, 2004, at [http://iaaweb.org/iaa/ Studies/nextsteps.pdf]; Roger A. Pielke, " NASA Needs a New Vision," *San Francisco Chronicle*, October 7, 2007, p. E-5, at [http://www. sfgate.com/cgi-bin/article.cgi?file=/c/a/2007/10/07/edbjshsmh.dtl]; Donald A. Beattie, "NASA's troubled future," The Space Report, September 4, 2007, at [http://www.thespacereview.com/article/945/1]; USA Today, "3 Alternate Visions for NASA's Future," April 11, 2005, at [http://www.usatoday. com/tech/science/space/ 2005-04-11-space-alternate-visions_x.htm].

[62] See earlier discussion for Senate and House Committee on Appropriations report language; also Lennard Fisk, Chair, Space Studies Board, National Research Council and Thomas M. Donahue Collegiate Professor of Space Science, University of Michigan, *The President's Vision for Space Exploration: Perspectives from a Recent NRC Workshop on National Space*

Policy, Testimony before the House Committee on Science, March 10, 2004, at [http://science.house.gov/Commdocs/hearings/ full04/mar10/ fisk.pdf].

In: Space Policy and Exploration
Editor: W. N. Callmers, pp. 39-62
ISBN: 978-1-60456-448-8

Chapter 2

SPACE LAUNCH VEHICLES: GOVERNMENT ACTIVITIES, COMMERCIAL COMPETITION, AND SATELLITE EXPORTS*

Carl E. Behrens

ABSTRACT

Launching satellites into orbit, once the exclusive domain of the U.S. and Soviet governments, today is an industry in which companies in the United States, Europe, China, Russia, Ukraine, Japan, and India compete. In the United States, the National Aeronautics and Space Administration (NASA) owns and launches its space shuttle. Private sector companies provide launch services for other NASA launches and most Department of Defense (DOD) launches. Commercial customers purchase launch services from the U.S. companies or their competitors. Since the early 1980s, Congress and successive Administrations have taken actions, including passing several laws, to facilitate the U.S. commercial space launch services business. The Federal Aviation Administration (FAA) regulates the industry.

Forecasts in the 1990s suggesting significant increases in launch demand sparked plans to develop new launch vehicles. NASA and DOD created government-industry partnerships to develop new reusable launch vehicles (RLVs) and "evolved" expendable launch vehicles (EELVs), respectively. (The space shuttle is the only RLV today. All other launch vehicles are "expendable" — they can only be used once). Several U.S. private sector

---

* Excerpted from CRS Report IB93062, dated March 20, 2006.

companies began developing their own launch vehicles. Projections for launch services demand declined dramatically beginning in 1999, however. NASA's efforts to develop a new RLV to replace the shuttle faltered. DOD's new EELVs (Atlas V and Delta IV) began service, but, with reduced demand, the companies that build them (Lockheed Martin and Boeing) want more DOD funding to defray their costs. In 2005, the two companies announced plans to merge their EELV launch services for U.S. government customers. The joint venture, if approved by regulatory authorities, would be named the United Launch Alliance. Commercial launch services would not be affected.

Congress is debating the future of the space shuttle, which returned to flight in July 2005 after a two and one-half year hiatus following the 2003 *Columbia* tragedy. President Bush directed NASA to terminate the shuttle in 2010, but some want it to continue until a replacement is available. NASA also is embarking on a program to build two new "shuttle-derived" launch vehicles to implement the President's "Vision for Space Exploration" to return astronauts to the Moon and someday send them to Mars. The FY2007 request includes $837 million for the first of these new vehicles, the Crew Launch Vehicle.

In October 2004, Burt Rutan's SpaceShipOne suborbital spacecraft won the $10 million Ansari X-prize. Some believe this heralds an era of comparatively affordable space tourism. Congress passed a law in 2004 (P.L. 108-492) that creates a regulatory environment for space tourism. Concerns that China benefitted militarily from knowledge gained through commercial satellite launches in the 1990s led to changes in U.S. satellite export policy. The changes, especially returning control over such exports to the State Department from the Commerce Department, remain controversial because of what some claim is a negative impact on U.S. satellite manufacturing companies whose clients may choose European suppliers to avoid the U.S. export control regulations.

Most Recent Developments

The next launch of the space shuttle is now targeted for July 2006, according to a NASA announcement March 14, 2006. Shuttle flights were indefinitely postponed after the STS-114 mission in July-August 2005 because of a foam-shedding event from the shuttle's External Tank that was similar to what led to the loss of the space shuttle *Columbia* in 2003. Hurricane damage to the Michoud Assembly Facility (MAF) in New Orleans, LA, where the External Tanks are manufactured, slowed the work needed to resume flights. As directed by President Bush, NASA plans to retire the shuttle in 2010. It hopes to conduct 16 or 17 more shuttle launches by then. The FY2007 budget request for the shuttle is $4.1

billion, compared with $4.8 billion allocated to the shuttle in NASA's FY2006 Initial Operating Plan. Over the 5-year "runout" (FY2007-2011), the request significantly augments the space shuttle account because NASA concluded in 2005 that it needed $3-5 billion more for the shuttle than was included in the agency's FY2006 5-year spending plan. The FY2007 request redirects $2 billion from NASA's science programs and $1.5 billion from its exploration programs to fund the shuttle, as well as additional money for the International Space Station. NASA's Stennis Space Center, MS, where the space shuttle's main engines are tested, also suffered hurricane damage. NASA estimates that it needs $760 million to cope with damage from Hurricane Katrina. The Administration requested $325 million, and Congress provided $350 million, for NASA for hurricane relief in the DOD appropriations and hurricane recovery act (P.L. 109-148). NASA is hoping that it will be able to request the remaining needed funds in a supplemental.

NASA's FY2007 request also includes $837 million to build a new launch vehicle, the Crew Launch Vehicle (CLV), to support President Bush's Vision for Space Exploration. The CLV is a "shuttle-derived" design. Furthermore, NASA is allocating $500 million through FY2009 to assist commercial companies develop and demonstrate launch vehicles that could service the International Space Station.

For FY2007, DOD is requesting $937 million for procurement of EELVs and $19 million for research and development (R and D).

Background and Analysis

U.S. Launch Vehicle Policy

The National Aeronautics and Space Administration (NASA), the Department of Defense (DOD), and the private sector have developed expendable launch vehicles (ELVs) — they can only be used once — to place satellites in orbit. NASA also developed the partially reusable space shuttle. U.S. ELVs currently in use include Atlas (manufactured by Lockheed Martin), Delta (Boeing), and Pegasus and Taurus (Orbital Sciences Corporation). Delta IV and Atlas V are the most recent additions to the fleet, and were developed through DOD's Evolved Expendable Launch Vehicle (EELV) program.

From "Shuttle-Only" to "Mixed Fleet"

In 1972, President Nixon approved NASA's plan to build the first reusable launch vehicle, called the space shuttle, and directed that it become the nation's primary launch vehicle, replacing all the ELVs except Scout (later discontinued for unrelated reasons). This would have made NASA and DOD dependent on a single launch vehicle, but the resulting high launch rate was expected to reduce the cost per flight significantly. The shuttle was first launched in 1981, and was declared operational in 1982. The phase-out of the ELVs began, but in 1984 the Air Force successfully argued that it needed a "complementary" ELV as a backup to the shuttle for "assured access to space" and initiated what is now known as the Titan IV program. Production lines for the Delta and Atlas began to close down, and it was expected that only the shuttle, Scouts, and Titan IVs would be in use by the mid-1980s.

Everything changed on January 28, 1986, however, when the space shuttle *Challenger* exploded 73 seconds after launch. Apart from the human tragedy, the *Challenger* accident deeply affected U.S. space launch policy, demonstrating the vulnerability of relying too heavily on a single system. Many military and civilian satellites had been designed to be launched on the shuttle, and could not have been transferred to ELVs even if the ELVs were not already being phased out. The remaining ELVs had their own problems in 1986. A Titan exploded in April and a Delta failed in May, which also grounded Atlas because of design similarities. Consequently, the Reagan Administration revised U.S. launch policy from primary dependence on the shuttle to a "mixed fleet" approach where a wide variety of launch vehicles are available. The shuttle is used principally for missions that require crew interaction, while ELVs are used for launching spacecraft. President Reagan also decided that commercial payloads could not be flown on the shuttle unless they were "shuttle-unique" (capable of being launched only by the shuttle or requiring crew interaction) or if there were foreign policy considerations. That action facilitated the emergence of a U.S. commercial space launch industry whose participants had long argued that they could not compete against government-subsidized shuttle launch prices. The White House and Congress had taken steps beginning in 1983 to assist in developing a commercial space launch services business, including President Reagan's 1983 designation of the Department of Transportation as the agency responsible for facilitating and regulating the commercial space launch sector. Passage of the 1984 Commercial Space Launch Act (P.L. 98-575), the Commercial Space Launch Act Amendments of 1988 (P.L. 100-657), and the Commercial Space Act of 1998

(P.L. 105-303) also have helped. But removing the shuttle as a competitor is seen as the major factor in fostering the U.S. commercial space launch business.

Clinton Administration Policy

On August 5, 1994, President Clinton released a National Space Transportation Policy that gave DOD lead responsibility for improving ELVs and NASA lead responsibility for upgrading the space shuttle and technology development of new reusable launch vehicles. The policy also set guidelines for the use of foreign launch systems, the use of excess ballistic missile assets for space launch, and encourages an expanded private sector role in space transportation R and D.

George W. Bush Administration Policy

On December 21, 2004, President Bush authorized a new U.S. Space Transportation Policy that supersedes the 1994 Clinton policy. A fact sheet on the Bush policy was released on January 6, 2005 [http://www.ostp.gov/html/SpaceTransFactSheetJan2005.pdf]. The new policy calls both for continued government support for space transportation capabilities, and for capitalizing on the U.S. private sector's "entrepreneurial spirit." DOD is directed to maintain the capability to develop, evolve, operate and purchase services for space transportation systems, infrastructure, and support activities necessary for national security requirements. NASA is directed to maintain the same capability for the civil sector, and to engage in development activities only for those requirements that cannot be met by capabilities being used by the national security or commercial sectors. The policy also directs NASA, in cooperation with DOD, to develop options to implement the President's January 2004 Vision for Space Exploration. NASA subsequently decided to build shuttle-derived launch vehicles. The policy states that the government will continue to support both EELVs (Boeing's Delta IV and Lockheed Martin's Atlas V) until the Secretary of Defense certifies to the President that U.S. assured access to space can be maintained without both providers.

U.S. LAUNCH VEHICLE PROGRAMS AND ISSUES

NASA's Space Shuttle Program

The Space Transportation System (STS) — the space shuttle — is a partially reusable launch vehicle and is the sole U.S. means for launching humans into orbit. It consists of an airplane-like Orbiter, with two Solid Rocket Boosters (SRBs) on each side, and a large, cylindrical External Tank (ET) that carries fuel for the Orbiter's main engines. The Orbiters and SRBs are reused; the ET is not. NASA has three remaining spaceflight-worthy Orbiters: *Discovery*, *Atlantis*, and *Endeavour*.

The Challenger and Columbia Tragedies

A total of 114 shuttle launches have taken place since April 1981. Two ended in tragedy, each killing seven astronauts. In 1986, the space shuttle *Challenger* exploded 73 seconds after launch because of the failure of a seal (an O-ring) between two segments of an SRB. In 2003, the space shuttle *Columbia* disintegrated as it returned to Earth after 16 days in orbit (see CRS Report RS21408). *Columbia* broke apart from aerodynamic forces after the left wing was deformed from the heat of gases that entered the wing through a hole caused during launch by a piece of foam insulation that detached from the ET. The *Columbia* Accident Investigation Board (CAIB) found that the tragedy was caused by technical and organizational failures, and made 29 recommendations, 15 of which it said should be completed before the shuttle returned to flight (see CRS Report RS21606 for a synopsis of the CAIB report).

Sean O'Keefe, NASA's Administrator from December 2001-February 2005, said NASA would comply with the CAIB recommendations. He established an RTF Task Group, chaired by two former astronauts, Tom Stafford and Dick Covey, to oversee NASA's implementation of the CAIB's 15 RTF recommendations. The Stafford/Covey Task Group [http://www.returntoflight.org] did not address management and culture changes, and was not tasked to determine whether the shuttle was ready to return to flight. Its assignment only was to evaluate NASA's compliance with the CAIB recommendations for RTF. The Task Group ultimately concluded that NASA met the intent of 12 of the 15 CAIB RTF recommendations, but not the other three: eliminating debris shedding from the External Tank, hardening the Orbiter so it can better survive debris impacts, and developing an on-orbit method of repairing the shuttle's thermal protection system. Dr. Michael Griffin, who became NASA Administrator in April 2005, said that NASA and contractor personnel are those responsible and accountable for determining if and when the shuttle is ready for RTF, and would not commit to meeting every CAIB recommendation.

Return to Flight (RTF)

NASA launched the space shuttle *Discovery* on the first of two "Return to Flight" (RTF) missions — STS-114 — on July 26, 2005 and it successfully landed on August 9. On July 27, however, NASA announced that a piece of foam had detached from STS-114's ET during launch, similar to what happened to *Columbia.* Cameras and other sensors on *Discovery* and on the International Space Station — to which *Discovery* was docked for much of its mission — imaged the Orbiter and determined that it was not damaged, but further shuttle launches have been suspended until the problem is resolved. Meanwhile, the images revealed that two "gapfillers" — ceramic coated fabric placed between thermal protection tiles — were protruding on the belly of the Orbiter that could have affected aerodynamic heating during reentry. One of the *Discovery* astronauts removed them during a space walk. The second RTF mission — STS-121 — was scheduled for September 2005, but now planned for July 2006.

The United Space Alliance (USA)

In 1995, NASA decided to turn most shuttle operations over to a "single prime contractor" — the United Space Alliance (USA), a limited liability company owned 50-50 by Boeing and Lockheed Martin. USA was created to pull together the 86 separate contracts with 56 different companies under which the shuttle program was then operating. NASA signed a $7 billion, six-year Space Flight Operations Contract (SFOC) with USA on September 26, 1996, with the goal of reducing shuttle operational costs while ensuring safety. The SFOC contract has been extended to 2006. NASA officials assert that SFOC has saved NASA $1 billion a year compared to what the costs would have been without it. Contracts for the External Tank, Solid Rocket Boosters (SRBs), and Space Shuttle Main Engines (SSMEs) have not been incorporated into SFOC. NASA manages those contracts: with Lockheed Martin for the External Tank; ATK Thiokol for the SRBs, and Pratt and Whitney Rocketdyne (owned by United Technologies) for the SSMEs.

The Shuttle's Future

As discussed below, NASA attempted unsuccessfully for many years to develop a "2nd generation" reusable launch vehicle (RLV) to replace the shuttle (which is the 1st generation RLV). In a November 2002 amendment to its FY2003 budget request, NASA announced a new space transportation strategy that indicated the shuttle would continue flying until at least 2015, and perhaps 2020 or beyond. The *Columbia* tragedy, and President Bush's January 2004 announcement of a new "Vision for Space Exploration" — to return astronauts to the Moon by 2020 and someday send them to Mars — forced NASA to revise that plan.

The President's Vision calls for the shuttle program, which absorbs approximately 25% of NASA's annual budget, to be terminated in 2010. A primary motivation is to make that funding available to implement other aspects of the Vision, although there also is concern about shuttle safety. Congress has been debating the Vision, including its impact on the shuttle and on U.S. human access to space. Some Members want to terminate the shuttle earlier than 2010 because they feel it is too risky and/or that the funds should be spent on accelerating the Vision. Others want to retain the shuttle at least until a new spacecraft, the Crew Exploration Vehicle (CEV), is available to take astronauts to and from the ISS. Under the plan announced by the President, CEV would not be ready at least until 2014, leaving a multi-year gap during which U.S. astronauts would have to rely on Russia for access to the ISS. NASA Administrator Griffin hopes to accelerate CEV development and have it ready by 2012, but has made clear that he views 2010 as a firm deadline for terminating the shuttle.

At the beginning of 2005, NASA officials indicated 28 shuttle flights were needed to complete ISS construction. When he became Administrator, Dr. Griffin stated that such a number was not achievable. In the fall of 2005, NASA announced a new plan showing 18 shuttle missions to the ISS, and possibly one more to service the Hubble Space Telescope (see CRS Report RS21767). As part of its FY2007 budget request, NASA reduced to 16 the number of shuttle flights it now plans to the ISS (plus one possible flight to Hubble). The two that were cut were logistics flights taking cargo to the ISS. NASA is hope that commercial launch service provider will develop vehicles that can perform that task. The original versions (H.R. 3070/S. 1281) of what became the 2005 NASA authorization act (P.L. 109-155) had conflicting language about the future of the shuttle. The original Senate bill directed NASA not to terminate the shuttle until a replacement was available; the House version directed NASA not to fly the shuttle after December 31, 2010. The final law states that it is U.S. policy to have human access to space on a continuous basis, and directs NASA to submit several related reports to Congress.

A related issue is what steps NASA and USA must take to make certain that the shuttle workforce retains the needed personnel and skills to ensure the shuttle flies safely through its remaining years of service. A 2005 GAO report (GAO-05-

230) concluded that NASA needs to better position itself to address future shuttle workforce needs.

Shuttle Budget

NASA's FY2006 Initial Operating Plan shows that NASA plans to spend $4.8 billion on the shuttle in FY2006. NASA Administrator Griffin conceded in 2005 that NASA's FY2006 budget request (prepared prior to his arrival at NASA) contained estimated ("placeholder") figures for shuttle funding in FY2008-2010 that were significantly lower than what is actually needed, and the program was underfunded by $3-5 billion over those three years. The FY2007 budget request adds $2 billion to the shuttle program to help compensate for that shortfall. For FY2007, NASA is requesting $4.1 billion for the shuttle program. The projected figures for FY2008-2010 are $4.1 billion, $3.8 billion, and $3.7 billion respectively, with a FY2011 projected funding level of $147 million as the program is terminated.

"Shuttle-Derived" Launch Vehicles

On September 19, 2005, NASA announced its implementation plan for the Vision for Space Exploration. Under the plan, NASA will develop two new launch vehicles that are derived from the space shuttle system. One is for launching the new Crew Exploration Vehicle (CEV); the other for launching large amounts of cargo. The first is designated the Crew Launch Vehicle (CLV), and would use a single SRB, augmented by a new "upper stage," with the CEV on top. It is informally referred to as the "single stick" design. The cargo launch vehicle, referred to as a "heavy lift" vehicle because it will be able to launch 125 tons of cargo to Earth orbit, would use a modified ET, supplemented by SRBs (like the shuttle), with the cargo spacecraft on top. NASA Administrator Griffin and Dr. Ronald Sega, Under Secretary of the Air Force, agreed that NASA would develop these two shuttle-derived vehicles for implementing the Vision, but use DOD's Evolved Expendable Launch Vehicles (EELVs) for its other spacecraft in the 5-20 metric ton range to the maximum extent possible (although new commercially-developed launch vehicles may compete with the EELVs if they become available).

NASA's Efforts to Develop New Reusable Launch Vehicles (RLVs)

U.S. expendable and reusable launch systems remain expensive and less efficient and reliable than desired. DOD and NASA initiated several efforts in the late 1980s and early 1990s to develop new systems, but each was terminated in turn because Congress or the agencies themselves were not convinced that the required investment had sufficient priority. In response to the 1994 Clinton policy, DOD's Evolved Expendable Launch Vehicle (EELV) program, and NASA's Reusable Launch Vehicle (RLV) program, were initiated.

Proponents believe that RLV technology can dramatically lower the cost of accessing space. NASA's efforts to develop a "2nd generation" RLV to replace the shuttle have not fared well, however. Starting in 1995, NASA pursued two "X" (for "experimental") flight test programs: X-33 and X-34. X-33 was a joint program with Lockheed Martin to build a subscale prototype of a large RLV based on single-stage-to-orbit (SSTO) technology. The SSTO concept involves a rocket that can attain orbit with only one stage (instead of two or more as is common today) carrying people or cargo. X-34 was a small RLV "testbed" to demonstrate reusable two-stage-to-orbit technologies, which was being built under a traditional contract with Orbital Sciences Corporation. NASA terminated X-33 and X-34 in March 2001 because the cost to complete them was too high relative to the benefits. NASA spent $1.2 billion on X-33, and Lockheed Martin said that it spent $356 million of its own funding. NASA spent $205 million on X-34.

NASA restructured its RLV program in 2000 (as part of its FY2001 budget request) and initiated the Space Launch Initiative (SLI). It then restructured the SLI program in 2002, and terminated it following President Bush's announcement of the Vision in January 2004. The goal was to develop RLV technology that would be "10 times safer and crew survivability 100 times greater, all at one-tenth the cost of today's space launch systems." The failure of the X-33 and X-34 programs, and of the National AeroSpace Plane (NASP) program before them, made some observers skeptical about NASA's ability to develop a 2nd generation RLV. In documentation accompanying a November 2002 budget amendment, NASA conceded that a new RLV lacked economic justification. SLI was restructured into two components: building an Orbital Space Plane (OSP), a spacecraft (not a launch vehicle) to take crews to and from the space station, and developing "Next Generation Launch Technology" (NGLT). Concurrent with President Bush's announcement of the Vision, NASA terminated SLI.

DOD's Evolved Expendable Launch Vehicle (EELV) Program

DOD began what is now known as the EELV program in FY1995 (P.L. 103-335) with a $30 million appropriation. EELV was first formally identified in

DOD's FY1996 budget. Two EELVs were developed in joint government-private sector programs: Boeing's Delta IV and Lockheed Martin's Atlas V. Both vehicles have successfully entered service (although the first launch of the Delta IV "heavy" — i.e., the version that can launch the greatest amount of mass — in December 2004 did not reach its intended orbit). The goal of the EELV program is to reduce launch costs by 25%.

In 1996, the Air Force selected Lockheed Martin and McDonnell Douglas (later bought by Boeing) for development contracts worth $60 million. Originally, one of those companies would have been selected in 1998 to develop the EELV. In November 1997, responding to indicators at the time that the commercial space launch market would be larger than expected, DOD announced that it would help fund development of both Atlas V and Delta IV. In October 1998, DOD awarded Boeing $1.88 billion for the Delta IV ($500 million for further development plus $1.38 billion for 19 launches), and awarded Lockheed Martin $1.15 billion for the Atlas V ($500 million for further development plus $650 million for 9 launches). The companies were expected to pay the rest of the development costs themselves. (Boeing officials state that Boeing invested $2.5 billion in design, development, and infrastructure for the Delta IV, of which the company wrote off $2 billion.)

In 2000, however, new market forecasts showed a reduction in expected commercial demand, and DOD reevaluated its EELV strategy. It renegotiated the contracts with both companies, including relieving Lockheed Martin (reportedly at the company's request) of the requirement to build a launch pad at Vandenberg AFB, CA. Each company built a launch pad for its vehicle at Cape Canaveral, FL for east coast launches. Both were expected to build them at Vandenberg for west coast launches (which launch site to use is determined by the type of orbit required by the satellite), but under this agreement, only Boeing would be able to launch from the west coast, giving it a monopoly on those EELV contracts. The companies also approached DOD to obtain additional government funding because of the downturn in the commercial market. This is called "assured access to space" in the sense of assuring that both companies remain in the EELV business so DOD has redundancy in capability should one of the launch vehicles experience difficulties. The FY2004 DOD authorization act (P.L. 108-136) codified "assured access" as U.S. policy.

Following revelations of ethics violations by Boeing in 2003, including some related to the EELV program, DOD suspended three Boeing business units from eligibility for new government contracts, shifted seven existing launch contracts from Boeing to Lockheed Martin, and disqualified Boeing from bidding for three new launch contracts. Boeing withdrew the Delta IV from competition for commercial contracts because it did not believe it could successfully compete. DOD reinstated the plan for an Atlas V launch pad at Vandenberg, so Boeing no longer would have a monopoly on west coast launches. The Boeing suspension was lifted on March 4, 2005.

The EELV program breached the "Nunn-McCurdy" limit of 25% cost growth, which requires DOD to cancel or restructure the program, or certify that it is essential to national security. In April 2004, DOD made that certification. In its report on the FY2005 DOD appropriations bill (H.Rept. 108-553), the House Appropriations Committee directed DOD to study whether both families of EELVs were really needed (H.Rept. 108-553). The December 2004 Bush space transportation policy directs DOD to continue to support both EELVs until the Secretary of Defense certifies to the President that assured access can be maintained without two EELV providers.

On May 2, 2005, Boeing and Lockheed Martin announced they planned to merge the production, engineering, test, and launch operations associated with EELV launch services for the U. S. government. If approved by U.S. and foreign regulatory bodies, both vehicles would be built at a single location (Boeing's facility in Decatur, AL), while engineering and administrative activities would be consolidated at Lockheed Martin's offices in Denver, CO. The joint venture, to be owned 50-50 by the two companies, would be called United Launch Alliance. The companies said the merger would save $100-150 million per year for the U.S. government. The merger does not affect commercial launch services.

For FY2007, DOD is requesting $937 million for procurement of EELVs and $19 million for research and development (R and D).

Private Sector Launch Vehicles (Including Space Tourism and the X-Prize)

Several entrepreneurial U.S. companies have been attempting to develop RLVs through private financing. Many have encountered difficulties in obtaining financing from the financial markets, and some have sought government loan guarantees or tax credits. Some have received limited direct government funding through various contracts. One company, SpaceX, headed by Elon Musk (creator of PayPal), asserts that it will dramatically reduce the cost of reaching orbit with its partially reusable Falcon launch vehicle. The first Falcon launch was scheduled for 2004, but has been delayed by a variety of factors.

Some companies are focusing on suborbital rockets instead of those that can attain orbit, anticipating that suborbital space tourism will be a substantial market.

Twenty seven teams from seven countries competed in the "Ansari X-Prize" contest [http://www.xprize.com] to win $10 million by becoming the first privately-financed company to launch a vehicle capable of carrying three people (one person actually had to be aboard) to an altitude of 100 kilometers (62.5 miles), return safely to Earth, and repeat it within two weeks using the same vehicle. On October 4, 2004, Burt Rutan's Scaled Composites Inc. won the X-prize with the SpaceShipOne vehicle, financed by Microsoft co-founder Paul Allen. Sir Richard Branson, head of the Virgin Group, is licensing the SpaceShipOne technology. He founded a company, Virgin Galactic, to offer commercial suborbital flights, and someday orbital flights. He reportedly expects to invest about $100 million in the new spaceships and associated ground infrastructure, and charge $200,000 per person per flight.

P.L. 108-492 creates a regulatory environment for commercial human space flight ("space tourism"). It sets requirements for protecting third parties, and for the crews of commercial spacecraft. There are few regulations for passengers ("spaceflight participants"), based on the philosophy that anyone who is willing to take the risk to fly on these new spacecraft should be allowed to do so as long as they are informed of the vehicle's safety record. If there are a significant number of accidents, or incidents that could have led to accidents, the FAA may set further passenger regulations. After eight years, the FAA may set any regulations it wishes. The FAA published a proposed rule in the December 29, 2005 *Federal Register* seeking public comment. Others believe that commercial human space flight should be regulated more strongly. H.R. 656 would strengthen safety requirements for passengers. A House Science subcommittee held a hearing on April 20, 2005, where Mr. Rutan sharply criticized FAA's regulatory process.

U.S. Commercial Launch Services Industry

Congressional Interest

Congress has been debating issues involving the domestic launch services industry for many years. Part of the debate has been focused on satellite export issues (discussed below). Another part concerns what the government should do to stimulate development of new launch vehicles by the private sector. That debate focuses on whether tax incentives or loan guarantees should be created for companies attempting to develop lower cost launch vehicles. Tax incentive advocates argue that loan guarantee programs allow the government to pick winners and losers; loan guarantee advocates argue that tax incentives are insufficient to promote necessary investment in capital intensive projects. Congress created (Title IX, FY2003 DOD appropriations Act, P.L. 107-248) a loan guarantee program for companies developing commercial, reusable, in-orbit space transportation system, but such systems are not launch vehicles (they move satellites from one orbit to another) and are not discussed further here. In 2004, Congress passed a law (P.L. 108-428) extending third-party liability indemnification for the commercial space launch industry for five more years. The government indemnifies commercial space launch companies for third-party claims between $500 million and $2 billion. Other 2004 legislation relating to regulation of space tourism is discussed above. In the 109th Congress, H.R. 3643 would make spaceports, like airports, eligible for tax exempt bonds.

Foreign Launch Competition

Europe, China, Russia, Ukraine, India, and Japan offer commercial launch services in competition with U.S. companies. Most satellites are manufactured by U.S. companies or include U.S. components and hence require export licenses, giving the United States considerable influence over how other countries participate in the commercial launch services market. The United States negotiated bilateral trade agreements with China, Russia, and Ukraine on "rules of the road" for participating in the market to ensure they did not offer unfair competition because of their non-market economies. Launch quotas were set in each of the agreements. However, President Clinton terminated the quotas for Russia and Ukraine in 2000, and the agreement with China expired at the end of 2001.

Europe

The European Space Agency (ESA) developed the Ariane family of launch vehicles. The first test launch of an Ariane was in 1979; operational launches began in 1982. ESA continued to develop new variants of Ariane. Ariane 5 is the only version now in use. ESA also is developing a smaller launch vehicle, Vega, whose first launch is expected in 2005. Operational launches are conducted by the French company Arianespace. Arianespace conducts its launches from Kourou, French Guiana, on the northern coast of South America. Arianespace also markets Russia's Soyuz launch vehicle and ESA is planning to build a launch site for Soyuz at Kourou. In 1985, a U.S. company (Transpace Carriers Inc.) filed an unfair trade practices complaint against Arianespace, asserting that European governments were unfairly subsidizing Ariane. The Office of the U.S. Trade

Representative (USTR) investigated and found that Europe was not behaving differently from the United States in pricing commercial launch services (then offered primarily on the government-owned space shuttle). The incident raised questions about what "rules of the road" to follow in pricing launch services. In the fall of 1990, USTR and Europe began talks to establish such rules of the road and assess how to respond to the entry of non-market economies into the launch services business. The only formal negotiating session was held in February 1991.

Each side is concerned about how much the respective governments subsidize commercial launch operations, but another controversial topic (not formally part of the talks) was whether Arianespace should be able to bid for launches of U.S. government satellites, which now must be launched on U.S. launch vehicles as a matter of U.S. policy. Arianespace wants that restriction lifted. France and other European governments do not have written policies requiring the use of Ariane for their government satellites. However, the member governments of ESA originally agreed to pay a surcharge of as much as 15-20% if they chose Ariane. The surcharge led some cost-conscious European governments to buy launch services from other (notably U.S.) suppliers. In the fall of 1995, ESA's member governments reached agreement with Arianespace to reduce the surcharge to encourage use of Ariane. ESA itself gives preference to using Ariane, but is not legally constrained from using other launch vehicles. Arianespace has encountered significant financial difficulties both because of the constrained market, and because of the failure of a new, more capable variant of the Ariane 5 in 2002. In May 2003, the ESA Council of Ministers adopted a European Guaranteed Access to Space (EGAS) program, providing 960 million euros for Arianespace to fix that variant of Ariane 5 (it successfully returned to flight in February 2005), and acquire Ariane 5 launch vehicles through 2009, while the commercial launch market is down.

China

The People's Republic of China offers several versions of its Long March launch vehicles commercially. China poses special issues not only because of its non-market economy, but because of technology transfer and political concerns. Launch services are offered through China Great Wall Industry Corp. (CGWIC). Because the United States currently will not issue export licenses for satellites or satellite components destined for China (see below), the Chinese commercial space launch program is sharply constrained.

U.S.-China Bilateral Trade Agreements for Launch Services

In 1989, China and the United States signed a six-year bilateral trade agreement restricting the number of Chinese commercial space launches to ensure China, with its nonmarket economy, did not unfairly compete with U.S. companies. A new seven-year agreement was reached in 1995, and amended in 1997. The agreement expired on December 31, 2001. While the agreements were in force, they established quotas on how many commercial satellites China could launch each year, and included pricing provisions to try to ensure that China did not unfairly compete with U.S. commercial launch service providers because of its non-market economy.

U.S. Satellite Exports to China: 1988-1997

In September 1988, the U.S. government agreed to grant three export licenses for satellites manufactured by Hughes to be launched by CGWIC. The Reagan Administration granted the licenses on the conditions that China sign three international treaties related to liability for satellite launches and other subjects; agree to price its launch services "on a par" with Western companies; and establish a government-to-government level regime for protecting technology from possible misuse or diversion. China met the conditions and the two countries signed a six-year agreement in January 1989. The now-defunct Coordinating Committee on Multilateral Export Controls (COCOM) approved the licenses that March.

On June 5, 1989, after the Tiananmen Square uprising, President George H. W. Bush suspended all military exports to China. At the time, exports of communications satellites were governed by the State Department's Munitions List. The satellites counted as military exports and the licenses were suspended. Then Congress passed language in the FY1990 Commerce, Justice, State and Judiciary appropriations (P.L. 101-162) and the 1990-91 Foreign Relations Authorization Act (P.L. 101-246, Section 902) prohibiting the export of U.S.-built satellites to China unless the President reported to Congress that (1) China had achieved certain political and human rights reforms, or (2) it was in the national interest of the United States. In December 1989, President Bush notified Congress that export of the satellites was in the national interest and the licenses were reinstated. The satellites were launched by China in 1990-1992.

A different issue arose in 1990. China signed a contract to launch an Arabsat Consortium satellite for $25 million, much less than what many considered "on a par" with Western companies. The main competitor, Arianespace, turned to both the French and U.S. governments to prohibit export of the satellite. No formal action was taken by the United States. In 1991, the Arabsat Consortium terminated the contract with the Chinese and signed an agreement with Arianespace, so the case became moot, but the issue of what constituted "on a par" remained. China argued that because its costs are so low, it could offer lower prices and still adhere to international norms as to what costs are included in setting the price. Yet another issue arose in 1991 — linkage of satellite export licenses with U.S. concern over China's ballistic missile proliferation policies. On April 30, 1991, the Bush Administration approved final export licenses for two satellites and for U.S. components of another, but to emphasize its concern about Chinese missile proliferation, disapproved export of U.S. components for a communications satellite China itself was building. On June 16, 1991, the White House announced that it would not approve any further export licenses for commercial satellite launches. On July 17, the State Department identified CGWIC as one of two Chinese entities engaged in missile technology proliferation activities that require the imposition of trade sanctions under the Arms Export Control Act, including denial of license applications for export items covered by the Missile Technology Control Regime (MTCR). Although the MTCR does not cover satellites (only satellite launch vehicles, which are close cousins of ballistic missiles), the identification of CGWIC as a cause of concern complicated China's marketing plans. China agreed to adhere to the MTCR, and the sanctions were lifted on February 21, 1992. In May 1992, INTELSAT agreed to launch at least one satellite on a Chinese launch vehicle. On September 11, 1992, the State Department notified Congress that it was waiving legislative restrictions on U.S. exports for six satellite projects with China. Many observers saw the move as a conciliatory gesture in the wake of the U.S. decision to sell F-16s to Taiwan.

On August 25, 1993, however, the U.S. government again imposed sanctions against China for ballistic missile proliferation activities, and the State Department said that satellite exports would not be permitted. The State Department announced October 4, 1994, it would lift the sanctions after China pledged to abide by the MTCR. During this period, tensions were acute between those viewing the sanctions as harmful to U.S. business interests and those seeking to prevent sensitive technology from reaching China and/or to punish China for MTCR infractions. The debate centered on whether the satellites should be governed by the State Department (Munitions List) or the Commerce Department (Commerce Control List). Some responsibility for export of commercial communications satellites was transferred from the State Department to the Commerce Department in 1992; in October 1996 primary responsibility was transferred to Commerce.

In January 1995, the launch of the Hughes-built APStar-2 satellite failed in-flight. Falling debris killed 6 and injured 23 on the ground. On February 6, 1996, President Clinton approved the export of four satellites to China for launch, despite concerns about China exporting nuclear weapons-related equipment to Pakistan. On February 14, 1996, a Long March 3B rocket carrying the INTELSAT 708 communications satellite built by Loral malfunctioned seconds after liftoff, impacting the ground and spreading debris and toxic fumes over the launch site and a nearby village. The Chinese reported 6 dead and 57 injured, but other reports suggested a higher figure. Some customers, including INTELSAT, canceled contracts. In May 1997, USTR stated that it believed China violated the pricing provisions of the bilateral agreement for the launching of Agila 2 for the Philippines. Chinese officials disagreed. On September 10, 1997, the *Washington Times* published a story that Chinese and Russian entities (including CGWIC) were selling missile technology to Iran. China denied the allegations.

Satellite Exports to China: 1998-2000 (Including the "Loral/Hughes" Issue, the Cox Committee Report, and Lockheed Martin)

On February 18, 1998, the President notified Congress that it was in the national interest to export Loral's Chinasat 8 satellite to China. On April 4, 1998, the *New York Times* reported that a 1997 classified DOD report alleged that Space Systems/Loral (part of Loral Space and Communications) and Hughes Electronics' satellite manufacturing division (then a subsidiary of General Motors; now Boeing Satellite Systems) provided technical information to China that improved the reliability of Chinese nuclear missiles. The assistance was provided in the wake of the February 1996 INTELSAT 708 launch failure (see above). The INTELSAT satellite was built by Loral, which participated in an inquiry into the accident at the request of insurance companies seeking assurances that the Chinese had correctly diagnosed and solved the cause of the failure. Loral formed a review committee that included representatives of other satellite companies, including Hughes. According to Loral, the review committee did not itself investigate the accident, but listened to Chinese officials explain their investigation and then wrote a report. Loral conceded that a copy of the report was given to the Chinese before it was provided to the State Department, in violation of Loral's internal policies. Loral says it notified the State Department when it learned that the Chinese had been given a copy. According to media sources, DOD's 1997 report says that the companies provided technical information in violation of Loral's export license. The companies insist they did not violate the licenses. The Justice Department investigated, and expanded the probe to include Hughes' response to the 1995 APStar-2 failure. A grand jury reportedly was empaneled in 1999. The government reached a civil settlement with Loral on January 9, 2002 where Loral agreed to pay a $14 million civil fine, and spend $6 million on strengthening its export compliance program. On December 26, 2002, the State Department charged Hughes Electronics and Boeing Satellite Systems with 123 export violations. The companies settled with the government on March 5, 2003, accepting a civil penalty of $20 million in cash, and $12 million in credits for money already spent ($4 million), or that will be spent ($8 million), on export program enhancements.

Many hearings on the "Loral/Hughes" issue were held by various House and Senate committees. In addition, the House established the Select Committee on U.S. National Security and Military/Commercial Concerns with the People's Republic of China, chaired by Representative Cox. The Cox committee concluded that Hughes and Loral deliberately transferred technical information and know-how to China during the course of accident investigations. The committee investigated other cases of China acquiring other U.S. technical information and made 38 recommendations.

The FY2000 DOD authorization act (P.L. 106-65) included language implementing many of the Cox committee recommendations. In brief, the Department of Justice must notify appropriate congressional committees when it is investigating alleged export violations in connection with commercial satellites or items on the munitions list if the violation is likely to cause significant harm or damage to national security with exceptions to protect national security or ongoing criminal investigations; companies must be provided with timely notice of the status of their export applications; enhanced participation by the intelligence community in export decisions is required; adequate resources must be provided for the offices at DOD and the State Department that approve export licenses; individuals providing security at overseas launch sites do not have to be DOD employees, but must report to a DOD launch monitor; and DOD must promulgate regulations concerning the qualifications and training for DOD space launch monitors and take other actions regarding those monitors and the records they maintain.

In February 1999, the Clinton Administration denied Hughes permission to export two satellites to China for launch. Export permission for those satellites (called APMTs) had been granted in 1997, but Hughes changed the spacecraft design, necessitating new export approval. That application was denied. On May 10, 2000, the White House made its first certification to Congress under the new process detailed in the FY1999 DOD authorization bill, approving the export to China of satellite fuels and separation systems for the Iridium program. On August 18, 2000, the State Department stated it would continue the suspension of a technical assistance agreement for Loral regarding launch of the Chinasat 8 satellite because the concerns that initiated the suspension in December 1998 had not been rectified. In January 2001, *Space News* reported that the Chinasat 8 export application was returned to Loral without action.

In April 2000, it became known that Lockheed Martin also was under investigation, in this case for performing a technical assessment, without an export license, of a Chinese "kick motor" used to place a satellite into its final orbit. On June 14, 2000, the State Department announced it had reached agreement with Lockheed Martin involving $13 million in penalties — $8 million that the company will pay over a four-year period and $5 million that was suspended and that the company can draw upon to fund a series of remedial compliance measures specified in the consent agreement.

Satellite Exports to China: 2001-Present

In July 2001, Senators Helms, Thompson, Shelby, and Kyl wrote to President Bush reportedly asking the President not to grant waivers for the export of satellites to China. As noted earlier, such waivers are required under the FY1990-91 Foreign Relations Authorization Act (P.L. 101-246). At the time, attention was focused on two European companies (Astrium and Alenia Spazio) that had built satellites for two multinational satellite organizations (INTELSAT and EUTELSAT, respectively) that were scheduled for launch by China. The satellites contain U.S. components. In August 2001, INTELSAT canceled its contract with Astrium for the APR-3 satellite, citing several factors, including the delay in obtaining U.S. export approval. EUTELSAT switched the launch of its satellite to Europe's Ariane. Other satellites being

manufactured by U.S. companies, however, such as Chinasat 8 and another being built by Loral (Apstar-5, for APT Satellite Co.), or containing U.S. components may require waivers in the future (see CRS Report 98-485 for a list of pending satellite exports). The FY2002 Commerce, Justice, State Appropriations Act (P.L. 107-77), the FY2003, FY2004, and FY2005 Consolidated Appropriations Acts (P.L. 108-7, 108-199, and 108-447), and the FY2006 Science, State, Justice, Commerce Appropriations Act (P.L. 109-108) require 15 days notice to Congress before processing licenses for exporting satellites to China.

Russia

U.S. policy prohibited U.S.-built satellites from being exported to the Soviet Union. Following the collapse of the Soviet Union, President George H. W. Bush said he would not oppose Russia launching an International Maritime Satellite

Organization (Inmarsat) satellite and the United States would negotiate with Russia over "rules of the road" for future commercial launches. Discussions in the fall of 1992 led to agreement in principle in May 1993; the agreement was signed on September 2, 1993, after Russia agreed to abide by the terms of the MTCR (see below). On January 30, 1996, the countries amended the agreement. Prior to Russia's first launch of a U.S.-built satellite, a Technology Safeguard Agreement among the United States, Russia, and Kazakstan (where the launch site is located) was signed in January 1999. A similar agreement for launches from Russia's Plesetsk, Svobodny, and Kapustin Yar launch sites was signed in January 2000.

The 1993 agreement was signed only after Russia agreed to comply with the MTCR in a case involving a Russian company, Glavkosmos, that planned to sell rocket engine technology to the Indian Space Research Organization (ISRO). The United States declared it violated the MTCR and imposed two-year sanctions against Glavkosmos and ISRO. In June 1993, the United States threatened to impose sanctions against Russian companies that did business with Glavkosmos. The two countries finally agreed that Russia would cease transferring rocket engine technology (the engines themselves were not at issue) to India. As noted, on September 10, 1997, the *Washington Times* published a story that Russian and Chinese entities, including the Russian Space Agency, were selling missile technology to Iran. In July 1998, Russia publicly identified nine entities that might be engaged in illegal export activities. The United States imposed sanctions against seven of them on July 28 and three more on January 12, 1999. The State Department said the United States would not increase the quota on geostationary launches that Russia could conduct under the 1996 agreement unless Russian entities ceased cooperating with Iran's ballistic missile program. The launches are conducted primarily by a U.S.-Russian joint venture composed of Lockheed Martin and Russia's Khrunichev and Energia, companies that were not among those sanctioned. Lockheed Martin was anxious to have the quota raised to 20 and eventually eliminated. On July 13, 1999, the White House raised the quota to 20, and eliminated it on December 1, 2000. (*Wall Street Journal,* December 1, 2000, p. A4).

Ukraine

Ukraine offers commercial launch services, chiefly as part of the Sea Launch joint venture among Boeing, Ukraine's Yuzhnoye, Russia's Energomash, and Norway's Kvaerner. The Sea Launch vehicle consists of a Ukranian two-stage Zenit rocket with a Russian third stage. The vehicle is launched from a mobile ocean oil rig built by Kvaerner. The rig is stationed in Long Beach, CA, where the launch vehicle and spacecraft are mated, and then towed into the ocean where the launch takes place. The United States and Ukraine signed a bilateral trade agreement in February 1996, that would have expired in 2001, but President Clinton terminated it on June 6, 2000, in recognition of "Ukraine's steadfast commitment to international nonproliferation norms." The first successful commercial launch was in October 1999. In 1998, Boeing agreed to pay $10 million for not abiding by export regulations in its dealings with Russia and Ukraine. Sea Launch announced plans in October 2003 to offer launches from Baikonur using Zenit beginning in 2005; the effort is called Land Launch.

India

India conducted its first successful orbital space launch in 1980. Its ASLV and PSLV launch vehicles can place relatively small satellites in low Earth orbit. India is developing a larger vehicle (GSLV) capable of reaching geostationary orbit. The GSLV, which uses Russian cryogenic engines that were the subject of a dispute between the United States and Russia (discussed earlier), made it first operational flight in September 2004.

Japan

Japan successfully conducted the first launch of its H-2 launch vehicle in 1994, the first all-Japanese rocket capable of putting satellites in geostationary orbit. Previous rockets used for this purpose were based on U.S. technology and a 1969 U.S.-Japan agreement prohibited Japan from launching for third parties without U.S. consent. With the H-2, Japan was freed from that constraint. H-2 was not cost effective, and encountered technical problems that led the Japanese government to abandon it in 1999. A new version, H2A, successfully completed its first launch in August 2001. In 2002, the Japanese government announced that it would privatize production of the H2A Mitsubishi Heavy Industries took over development and marketing. H-2A launches are conducted from Tanegashima, on an island south of Tokyo. In June 1997, the Japanese government reached agreement with the fishing industry to allow more launches from Tanegashima. Fishermen must evacuate the area near the launch site during launches. The agreement extends from 90 to 190 the number of days per year that launches may be conducted, and permits up to eight launches a year instead of two.

Satellite Exports: Agency Jurisdiction and Other Issues

Between 1992 and 1996, the George H. W. Bush and Clinton Administrations transferred responsibility for decisions regarding export of commercial satellites from the State Department to the Commerce Department. A January 1997 GAO report (GAO/NSIAD-97-24) examines that decision. In response to concerns about the launch of satellites by China (discussed above), Congress directed in the FY1999 DOD authorization bill (P.L. 105-261) that export control responsibility be returned to the State Department effective March 15, 1999. Which agency should control these exports remains controversial.

The 108th Congress debated, but did not clear, legislation on this topic. The House International Relations Committee (HIRC) reported H.R. 1950 (H.Rept. 108-105, Pt. 1), the FY2004 State Department Authorization Act, with language that would have left the decision on agency jurisdiction to the President if the export was to a NATO country or major non-NATO ally. Exports to China would have remained under State Department jurisdiction. The House Armed Services Committee rejected the HIRC language in its markup of the bill (H.Rept. 108-105, Pt. 3), however, and the House-passed version did not include that language. There was no further legislative action on that bill.

Some of the controversy reflects concerns of the aerospace and space insurance industries that the regulations are being implemented too broadly and vigorously. DOD officials and others have cited potential harm to the U.S. defense industrial base if U.S. exports are stifled, too. One concern is the length of time needed to obtain State Department approval. Section 309 of the FY2000 State Department authorization act (incorporated into the FY2000 Consolidated Appropriations Act, P.L. 106-113) directed the Secretary of State to establish an export regime with expedited approval for exports to NATO allies and major non-NATO allies. The rules took effect July 1, 2000. In May 2000, the State Department reportedly notified France that it would not apply strict technology export control on satellites to be launched by Ariane (*Space News*, May 29, 2000, p. 1). The Security Assistance Act (P.L. 106-280) reduced from 30 days to15 days the time Congress has to review decisions on exporting commercial communications satellites to Russia, Ukraine, and Kazakhstan, making the time period the same as for NATO allies.

The Satellite Industry Association (SIA) released figures in May 2001 showing U.S. satellite manufacturers losing market share to foreign companies. SIA and others attributed that loss in part to the shift in jurisdiction to State, which they assert creates uncertainty for satellite customers over when and whether export licenses will be approved. The trade publication *Space News* reports on the number of new commercial satellite orders awarded world-wide each year. According to that source, U.S. companies won 19 of the 22 contracts in 2001; three of the four in 2002; nine of the 16 in 2003; and nine of the 12 in 2004.

LEGISLATION

109th Congress, 1st Session

P.L. 109-108, H.R. 2862. FY2006 Science, State, Justice, Commerce appropriations bill (includes NASA). Reported from House Appropriations Committee June 10, 2005 (H.Rept. 109-118); passed House June 16. Reported from Senate Appropriations Committee June 23 (S.Rept. 109-88); passed Senate September 15. Conference report (H.Rept. 109-272) passed House November 9, Senate November 16. Signed into law November 22, 2005.

P.L. 109-148, H.R. 2863. FY2006 DOD appropriations bill. Reported from House Appropriations Committee June 10, 2005 (H.Rept. 109-119); passed House June 20. Reported from Senate Appropriations Committee September 29 (S.Rept. 109-141); passed Senate October 7. Conference report (H.Rept. 109-359) passed House December 19, Senate December 21 (amended). Signed into law December 30, 2005.

P.L. 109-155, S. 1281. FY2007-2008 NASA authorization bill. H.R. 3070 would have authorized NASA funding for FY2006-2007; reported by the House Science Committee July 18 (H.Rept. 109-173), passed House July 22, 2005. S. 1281, as originally passed, authorized NASA funding for FY2006-2010; reported by Senate Commerce Committee July 26 (S.Rept. 109-108), passed Senate September 28, 2005. Conference report (H.Rept. 109-354), which authorizes funding for FY2007-2008, passed House December 17, Senate December 21, 2005. Signed into law December 30, 2005.

P.L. 109-163, H.R. 1815. FY2006 DOD authorization bill. Reported from House Armed Services Committee May 20, 2005, H.Rept. 109-89; passed House May 25. S. 1042 reported from Senate Armed Services Committee May 17 (S.Rept. 109-69); passed Senate November 15. Conference report (H.Rept. 109-360) passed House December 19, Senate December 21. Signed into law January 6, 2006.

In: Space Policy and Exploration
Editor: W. N. Callmers, pp. 63-85
ISBN: 978-1-60456-448-8

Chapter 3

SPACE STATIONS*

Marcia S. Smith

ABSTRACT

Congress continues to debate NASA's International Space Station (ISS), a permanently occupied facility in Earth orbit where astronauts live and conduct research. Congress appropriated approximately $35 billion for the program from FY1985-2005. The initial FY2006 ISS request was $2.180 billion: $1.857 billion for construction and operations and $324 million for research to be conducted by ISS crews. In a July budget amendment, NASA transferred $168 million for ISS Crew/Cargo Services to another part of the NASA budget and reduced the ISS request commensurately. The final version of the FY2006 appropriations bill that includes NASA (H.R. 2862) cuts $80 million from the originally submitted budget, and NASA now plans to spend $306 million, instead of $324 million, on ISS research in FY2006.

The space station is being assembled in Earth orbit. ISS segments, crews, and cargo are taken into orbit by Russian or U.S. spacecraft. ISS has been permanently occupied by successive "Expedition" crews rotating on 4-6 month shifts since November 2000. "Expedition 12" is now aboard. Cost growth and schedule delays have characterized the program since its inception. The grounding of the space shuttle fleet after the 2003 *Columbia* tragedy and the July 2005 *Discovery* "Return to Flight" mission is further affecting schedule, operations, and cost. Most of the remaining ISS segments are designed to be launched by the shuttle and construction therefore is suspended. President Bush's January 2004 "Vision for Space Exploration" also is affecting the ISS program. He directed that the shuttle program be

---

* Excerpted from CRS Report IB93017, dated November 17, 2005.

terminated in 2010, and changed the focus of ISS-based research to only that which supports his "Moon/Mars" Vision instead of the broadlybased program that was planned.

Canada, Japan, and several European countries became partners with NASA in building the space station in 1988; Russia joined in 1993. Except for money paid to Russia, there is no exchange of funds among the partners. Europe, Canada, and Japan collectively expect to spend about $11 billion of their own money. A reliable figure for Russian expenditures is not available.

In 1993, when the current space station design was adopted, NASA said it would cost $17.4 billion for construction (not including launch or other costs). That estimate grew to $24.1-$26.4 billion, leading Congress to legislate spending caps on part of the program in 2000. The estimate then grew by almost another $5 billion, leading NASA (at White House direction) to cancel or indefinitely defer some hardware to stay within the cap. NASA exceeded the cap in FY2005, however.

Controversial since the program began in 1984, the space station has been repeatedly redesigned and rescheduled, often for costgrowth reasons. Congress has been concerned about the space station for that and other reasons. Twenty-two attempts to terminate the program in NASA funding bills were defeated, however (3 in the 106th Congress, 4 in the 105th Congress, 5 in the 104th, 5 in the 103rd, and 5 in the 102nd). Three other attempts in broader legislation in the 103rd Congress also failed.

Current congressional debate focuses on the impact of space shuttle-related delays, and the future of ISS in light of President Bush's new exploration initiative.

MOST RECENT DEVELOPMENTS

The "Expedition 12" crew, American John McArthur and Russian Valery Tokarev, continue their work aboard the International Space Station (ISS). They arrived via the Russian Soyuz TMA-7 spacecraft on October 3. They replaced the Expedition 11 crew, which returned to Earth in the Soyuz TMA-6 spacecraft on October 10. ISS crews are taken to and from the ISS on Russian Soyuz spacecraft while launches of the U.S. space shuttle are grounded. Shuttle launches have been postponed until at least May 2006 because of a problem during the most recent launch, STS-114, in July 2005 (see CRS Report RS21408). Until then, U.S. astronauts will remain dependent on Russia for access to the ISS. Russia has been providing crew transport and "crew return" (i.e., a "lifeboat" capability for emergencies) services to NASA at no cost under a 1996 agreement. Russia's obligations under that agreement have been fulfilled with the launch of Soyuz TMA-7. It will remain docked with the ISS during the Expedition 12 mission, scheduled to end in April 2006. In the future, NASA will have to pay for Soyuz services, but has been prohibited from doing so unless Russia complies with the Iran Nonproliferation Act (INA). The House and Senate have passed a bill (S. 1713) that would amend the INA to allow NASA to pay for such services through January 1, 2012 (see CRS Report RS22270)

The original FY2006 request for the ISS was $2.180 billion: $1.857 billion for construction and operations (including $160 million for ISS Crew/Cargo Services) and $324 million for research. In the final version of the FY2006 Science, State, Justice and Commerce (SSJC) appropriations bill (H.R. 2862), which includes NASA, Congress cut $80 million from the ISS program, including $60 million from ISS Crew/Cargo Services. The House passed a FY2006-2007 NASA authorization bill (H.R. 3070, H.Rept. 109-173); it does not specify an amount for the ISS program. The Senate passed a FY2006-2010 NASA authorization bill (S. 1281, S.Rept. 109-103) that adds $100 million for FY2006 and calls for enhancing the use of ISS for research. On July 15, NASA amended its FY2006 request, shifting the funds for ISS Crew/Cargo services (which it now identifies as $168 million) into a different budget account, and commensurately reducing the ISS request. The conference report on H.R 2862 does not reflect the budget amendment. See CRS Report RL32988 for details on the FY2006 NASA budget request.

BACKGROUND AND ANALYSIS

Introduction

NASA launched its first space station, Skylab, in 1973. Three crews were sent to live and work there in 1973-74. It remained in orbit, unoccupied, until it reentered Earth's atmosphere in July 1979, disintegrating over Australia and the Indian Ocean. Skylab was never intended to be permanently occupied. The goal of a permanently occupied space station with crews rotating on a regular basis was high on NASA's list for the post-Apollo years. In 1969, Vice President Agnew's Space Task Group recommended a permanent space station and a reusable space transportation system (the space shuttle) to service it as the core of NASA's program in the 1970s and 1980s. Budget constraints forced NASA to choose to build the space shuttle first. When NASA declared the shuttle "operational" in 1982, it was ready to initiate the space station program.

In his January 25, 1984 State of the Union address, President Reagan directed NASA to develop a permanently occupied space station within a decade, and to invite other countries to join. On July 20, 1989, the 20th anniversary of the first Apollo landing on the Moon, President George H. W. Bush gave a major space policy address in which he voiced his support for the space station as the cornerstone of a long-range civilian space program eventually leading to bases on the Moon and Mars. That "Moon/Mars" program, the Space Exploration Initiative, was not greeted with enthusiasm in Congress, primarily due to budget concerns, and ended in FY1993, although the space station program continued.

President Clinton dramatically changed the character of the space station program in 1993 by adding Russia as a partner to this already international endeavor. That decision made the space station part of the U.S. foreign policy agenda to encourage Russia to abide by agreements to stop the proliferation of ballistic missile technology, and to support Russia economically and politically as it transitioned from the Soviet era. The Clinton Administration strongly supported the space station within certain budget limits.

President George W. Bush, prompted in part by the February 2003 space shuttle *Columbia* tragedy, made a major space policy address on January 14, 2004, directing NASA to focus its activities on returning humans to the Moon and someday sending them to Mars. Included in this "Vision for Space Exploration" is a plan to retire the space shuttle in 2010. The President said the United States would fulfill its commitments to its space station partners, but the details of how to accomplish that without the shuttle were not announced.

THE SPACE STATION PROGRAM: 1984-1993

NASA began the current program to build a space station in 1984 (FY1985). In 1988, the space station was named *Freedom*. Following a major redesign in 1993, NASA announced that the *Freedom* program had ended and a new program begun, though NASA asserts that 75% of the design of the "new" station is from *Freedom*. The new program is simply referred to as the International Space Station (ISS). Individual ISS modules have various names. (Some refer to the facility as "Space Station Alpha," but that is not its formal name). ISS is a laboratory in space for conducting experiments in near-zero gravity ("microgravity"). A broadly based research program was planned, but President Bush's Vision for Space Exploration would limit U.S. research on ISS to that which is needed to support the goal of sending humans back to the Moon and to Mars. From FY1985- FY2005, Congress appropriated approximately $35 billion for the space station program.

Space Station *Freedom*

When NASA began the space station program in 1984, it said the program would cost $8 billion (FY1984 dollars) for research and development (R and D — essentially the cost for building the station without launch costs) through completion of assembly. From FY1985-1993, Congress appropriated $11.4 billion to NASA for the *Freedom* program. Most of the funding went for designing and redesigning the station over those years. Little hardware was built and none was launched. Several major redesigns were made. A 1991 redesign evoked concerns about the amount of science that could be conducted on the scaled-down space station. Both the White House Office of Science and Technology Policy (OSTP) and the Space Studies Board (SSB) of the National Research Council concluded that materials science research could not justify building the space station, and questioned how much life sciences research could be supported. They criticized the lack of firm plans for flying a centrifuge, considered essential to this research. NASA subsequently agreed to launch a centrifuge, but, in September 2005, decided that it was not needed (see below).

Cost estimates for *Freedom* varied widely depending on when they were made and what was included. *Freedom* was designed to be operated for 30 years. As the program ended in 1993, NASA's estimate was $90 billion (current dollars): $30 billion through the end of construction, plus $60 billion to operate it for 30 years. The General Accounting Office (GAO) estimated the total cost at $118 billion, including 30 years of operations.

In 1988, after three years of negotiations, Japan, Canada and nine European countries under the aegis of the European Space Agency (ESA) agreed to be partners in the space station program. A government-to-government Intergovernmental Agreement (IGA) was signed in September, and Memoranda of Understanding (MOUs) between NASA and its counterpart agencies were signed then or in 1989. The partners agreed to provide hardware for the space station at their own expense, a total of $8 billion at the time.

1993 Redesign — the Clinton Administration Restructuring

In early 1993, as President Clinton took office, NASA revealed $1 billion in cost growth on the *Freedom* program. The President gave NASA 90 days to develop a new, less costly, design with a reduced operational period of 10 years. A new design, *Alpha,* emerged on September 7, 1993, which NASA estimated would cost $19.4 billion. It would have used some hardware bought from Russia, but Russia was not envisioned as a partner. Five days earlier, however, the White House announced it had reached preliminary agreement with Russia to build a joint space station. Now called the International Space Station (ISS), it superseded the September 7 *Alpha* design. NASA asserted it would be a more capable space station and be ready sooner at less cost to the United States. Compared with the September 7 *Alpha* design, ISS was to be completed one year earlier, have 25% more usable volume, 42.5 kilowatts more electrical power, and accommodate six instead of four crew members.

In 1993, President Clinton pledged to request $10.5 billion ($2.1 billion a year) for FY1994-1998. NASA said the new station would cost $17.4 billion to build, not including money already expended on the *Freedom* program. That estimate was derived from the $19.4 billion estimate for the September 7 *Alpha* design minus $2 billion that NASA said would be saved by having Russia in the program. The $2.1 billion and $17.4 billion figures became known as "caps," though they were not set in law. (See Cost Caps below).

The International Space Station (ISS): 1993-Present

The International Space Station program thus began in 1993, with Russia joining the United States, Europe, Japan, and Canada. The 1993 and subsequent agreements with Russia established three phases of space station cooperation and the payment to Russia of $400 million, which grew to $473 million. (NASA transferred about $800 million to Russia for space station cooperation through this and other contracts.)

During Phase I (1995-1998), seven U.S. astronauts remained on Russia's space station *Mir* for long duration (several month) missions with Russian cosmonauts, Russian cosmonauts flew on the U.S. space shuttle seven times, and nine space shuttle missions docked with *Mir* to exchange crews and deliver supplies. Repeated system failures and two life-threatening emergencies on *Mir* in 1997 raised questions about whether NASA should leave more astronauts on *Mir*, but NASA decided *Mir* was sufficiently safe to continue the program. (As discussed below, *Mir* was deorbited in 2001.) Phases II and III involve construction of the International Space Station itself, and blend into each other. Phase II began in 1998 and was completed in July 2001; Phase III is underway.

ISS Design, Cost, Schedule, and Lifetime

ISS is being built by a partnership among the United States, Russia, Europe, Japan, and Canada. The 1988 Intergovernmental Agreement was renegotiated after Russia joined the program. The new version was signed in 1998. The IGA is a treaty in all the countries except the United States, where it is an Executive Agreement. The IGA is implemented through Memoranda of Understanding (MOUs) between NASA and its counterpart agencies. Brazil is not a partner in ISS, but agreed to participate through a bilateral agreement with NASA. Boeing is the U.S. prime contractor.

NASA originally stated that ISS would be operated for 10 years after assembly was completed, with a possibility for 5 additional years if the research was considered worthwhile. Using the original schedule, assembly would have been completed in 2002, with operations at least through 2012. The completion of assembly slipped to 2006, but President Bush restructured the space station program in 2001, and it was not clear when assembly would be "completed." NASA briefing charts in March 2003 showed space station operations possibly continuing until 2022. Under President Bush's January 2004 "Vision for Space Exploration," however, NASA plans to complete its utilization of ISS in 2016 (though the other partners may continue to use it after that time).

ISS segments are launched into space on U.S. or Russian launch vehicles and assembled in orbit. The space station is composed of a multitude of modules, solar arrays to generate electricity, remote manipulator systems, and other elements that are too numerous to describe here. Details can be found at [http://spaceflight.nasa.gov/home/index.html]. Six major modules are now in orbit. The first two were launched in 1998: Zarya ("Sunrise," a Russian-built, U.S.-owned, module with guidance, navigation, and control systems) and Unity (a U.S. "node" connecting other modules). Next was Zvezda ("Star," a Russian module that serves as the crew's living quarters) in 2000. Destiny (a U.S. laboratory), Quest (a U.S. airlock), and Pirs ("Pier," a Russian docking compartment) arrived in 2001. Among the other modules awaiting launch are laboratory modules built by Russia, Europe, and Japan, and two more "nodes" built by Europe. (Zarya counts as a U.S. module because NASA paid Russia to build it. Some of the European- and Japanese-built hardware counts as U.S. because they are built under barter agreements where Europe and Japan produce hardware NASA needs instead of paying cash to NASA for launch and other ISS-related services.) Ordinarily, the U.S. space shuttle takes crews and cargo back and forth to ISS. The shuttle system is currently grounded because of problems that occurred during the July 2005 launch of STS-114 (see CRS Report

RS21408). Russian Soyuz spacecraft are also used to take crews to and from ISS, and Russian Progress spacecraft deliver cargo, but cannot return anything to Earth (Progress is not designed to survive reentry into the Earth's atmosphere). A Soyuz is always attached to the station as a lifeboat in case of an emergency.

The schedule for launching segments and crews is called the "assembly sequence" and has been revised many times. At the end of the Clinton Administration, the assembly sequence showed completion of assembly ("assembly complete") in April 2006. The most recent assembly sequence was released after a January 2005 "Heads of Agency" meeting in Montreal, but it does not include launch dates, only the order in which the launches will go. It does list "Establishment of a Permanent Crew of Six (January 2009)," followed by nine shuttle launches to assembly complete. Under the Vision, ISS construction is to be completed by 2010, but NASA Administrator Griffin has indicated that a sufficient number of shuttle flights may not be able to be launched in that time period. He intends to terminate the shuttle in 2010 nonetheless, however, and reportedly is assessing other methods for launching ISS segments.

"Expedition" crews have occupied ISS on a 4-6 month rotating basis since November 2000. Originally the crews had three members (two Russians and one American, or two Americans and one Russian), with an expectation that crew size would grow to six or seven once assembly was completed. Crew size is temporarily reduced to two (one American, one Russian) while the U.S. shuttle is grounded in order to reduce resupply requirements. The number of astronauts who can live on the space station is limited in part by how many can be returned to Earth in an emergency by lifeboats docked to the station. Only Russian Soyuz spacecraft are available as lifeboats. Each Soyuz can hold three people, limiting crew size to three if only one Soyuz is attached. NASA planned to build a U.S. Crew Return Vehicle (CRV) to provide lifeboat capabilities for at least four more crew. The Bush Administration canceled those plans due to cost growth in the ISS program, then began a different program (the Orbital Space Plane) that also was cancelled. In September 2005, NASA announced that the new Crew Exploration Vehicle (CEV) it is building to implement the President's Vision for Space Exploration (the "Moon/Mars" program) will be designed to take crews to and from the ISS, and to serve as a lifeboat. NASA currently hopes to have it ready by 2012.

Each Soyuz must be replaced every six months. The replacement missions are called "taxi" flights since the crews bring a new Soyuz up to ISS and bring the old one back to Earth. Therefore, under normal conditions, the long duration Expedition crews are regularly visited by taxi crews, and by the space shuttle bringing up additional ISS segments or exchanging Expedition crews. When the shuttle is unavailable, Expedition crews are taken back and forth on the "taxi" flights.

September 1993-January 2001: The Clinton Administration

Cost Growth

From FY1994-FY2001, the cost estimate for building ISS grew from $17.4 billion to $24.1-26.4 billion. The $17.4 billion estimate (called its "development cost," "construction cost," or "R and D cost") covered FY1994 through completion of assembly, then scheduled for June 2002. It did not include launch costs, operational costs after completion of assembly, civil service costs, or other costs. NASA estimated the program's life-cycle cost (all costs, including funding spent prior to 1993) from FY1985 through FY2012 at $72.3 billion. In 1998, GAO estimated the life-cycle cost at $95.6 billion (GAO/NSIAD-98-147). More recent, comparable, life-cycle estimates are not available from NASA or GAO.

Cost growth first emerged publicly in March 1996 when then-NASA Administrator Daniel Goldin gave the space station program manager control of money allocated for (and previously overseen by) the science offices at NASA for research intended to be conducted aboard the space station. Congress gave NASA approval to transfer $177 million from those science accounts to space station construction in the FY1997 VA-HUD-IA appropriations act (P.L. 104-204). A similar transfer was approved for FY1996 ($50 million). NASA changed its accounting methods so future transfers would not require congressional action, and transferred $235 million from space station science into construction in FY1998. ("Space station science" funding is for scientific activities aboard the space station. It is separate from NASA's other "space science" funding, such as Mars exploration.)

One factor in the cost growth was schedule slippage related to Russia's Zvezda module. As insurance against further Zvezda delays, or a launch or docking failure, NASA decided to build an "Interim Control Module" (ICM). To cover cost growth associated with the schedule delay and ICM, NASA requested permission to move $200 million in FY1997 from the space shuttle and payload utilization and operations accounts to the space station program, and to transfer $100 million in FY1998 from unidentified NASA programs to the space station program. The appropriations committees approved transferring the $200 million in FY1997, but not the FY1998 funding.

In March 1998, NASA announced that the estimate for building the space station had grown from $17.4 billion to $21.3 billion. In April 1998, an independent task force concluded that the space station's cost through assembly complete could be $24.7 billion. Mr. Goldin initially refused to endorse the $24.7 billion estimate, but by 2000, NASA's own estimate had grown to $24.1-$26.4 billion.

Cost Caps

The $2.1 billion per year figure the Clinton White House and Congress agreed to spend on the space station, and NASA's $17.4 billion estimate to build the station, became known as "caps," although they were not set in law. Both were exceeded in 1997-1998. As costs continued to rise, Congress voted to legislate caps on certain parts of the ISS program in the FY2000-2002 NASA authorization act (P.L. 106-391). The caps are $25 billion for development, plus $17.7 billion for associated shuttle launches. The act also authorizes an additional $5 billion for development and $3.5 billion for associated shuttle launches in case of specified contingencies. The caps do not apply to operations, research, or crew return activities after the space station is "substantially" complete, defined as when development costs consume 5% or less of the annual space station budget. In its FY2006 budget justification (p. EC 2-4), NASA alerted Congress that it might exceed the $25 billion cap for ISS development during FY2005, attributing the increased costs to delays resulting from the *Columbia* tragedy. The House version of the pending NASA authorization bill, H.R. 3070, would repeal the cap. The Senate version, S. 1281, would require NASA to submit a report on how certain factors affected ISS development costs, and to identify statutory changes needed to address those impacts. NASA did exceed the cap in FY2005, having spent $25.7 billion in funds that are counted against the cap.

2001-Present: The George W. Bush Administration

Cost Growth

As President Bush took office, NASA revealed substantial additional cost growth. In 2000, NASA's estimate of the remaining cost to build ISS was $8 billion (FY2002 to FY2006). In January 2001, however, it announced that an additional $4.02 billion was needed. That figure grew to $4.8 billion by June, and the IMCE task force (discussed below) said another $366 million in growth was discovered between August and October. Those increases would have raised the cost to over $30 billion, 72% above the 1993 estimate, and $5 billion above the legislated cap. NASA explained that program managers had underestimated the complexity of building and operating the station. The Bush Administration signaled it supported the legislated cap, would not provide additional funds, and

NASA would have to find what it needed from within its Human Space Flight account.

"Core Complete" Configuration

In February 2001, the Bush Administration announced it would cancel or defer some ISS hardware to stay within the cap and control space station costs. It canceled the Propulsion Module, and "indefinitely deferred" the Habitation Module, Node 3, and the Crew Return Vehicle (CRV). The decision truncated construction of the space station at a stage the Administration called "core complete." The Administration said that "enhancements" to the station might be possible if NASA demonstrated improved cost estimating and program management. In 2001, the space station program office at Johnson Space Center (JSC) estimated that it would cost $8.3 billion from FY2002-2006 to build the core complete configuration, described at that time as all the U.S. hardware planned for launch through "Node 2," plus the launch of laboratories being built by Europe and Japan. NASA subsequently began distinguishing between "U.S. Core Complete" (the launches through Node 2, which, prior to the *Columbia* tragedy, was scheduled for February 2004) and "International Partner (IP) Core Complete" which included the addition of European and Japanese laboratory modules (then anticipated in 2008).

The non-U.S. partners, and U.S. scientists who planned to conduct research on ISS, expressed deep concern with the core complete configuration (see CRS Report RL31216). A major issue was that NASA reduced its space station research budget by 37.5% over the FY2002-2006 period, necessitating a reassessment of U.S. research priorities on ISS. A July 2002 report of the "Research Maximization and Prioritization" (ReMaP) task force, and a September 2002 National Research Council report, made recommendations on research priorities. Both were superseded by President Bush's January 2004 "Vision for Space Exploration," which directs that U.S research on ISS be restricted only to that which supports the Vision. A new research plan has not been released by NASA. At the time the core complete configuration was announced, another major concern was the decision to indefinitely defer the CRV, which subsequently was canceled. That would have limited the space station to three permanent crew members, not seven as planned, reducing the number of researchers on board to conduct the research program. As discussed elsewhere in this report, NASA reinstated plans for a CRV capability in 2005.

Reviews of NASA's Cost Estimates and Adding Funds for ISS

NASA created the ISS Management and Cost Evaluation (IMCE) Task Force in July 2001 to review the space station program office's $8.3 billion cost estimate for finishing the core complete configuration. Chaired by former Lockheed Martin executive Tom Young, IMCE determined that the cost estimate was not credible, and NASA should make significant management and cost estimating changes (see CRS Report RL31216). NASA Headquarters directed the space station program office to reassess its estimate, and had two independent groups conduct their own estimates. A July 2002 GAO report (GAO-02-735) concluded that NASA's focus on managing annual budgets resulted in NASA's failure to heed indicators of future program cost growth.

In November 2002, the Bush Administration submitted an amended FY2003 budget request that shifted $706 million into the ISS program for FY2004-2007: $660 million to boost program reserves, and $46 million in FY2004 for "long-lead" items to preserve the option of increasing crew size beyond three. (Congress cut $200 million from ISS in FY2004, however.) The latter included a proposal to build an Orbital Space Plane (OSP) to takes crews back and forth to ISS as a complement to the space shuttle.

At a December 2002 "Heads of Agency" meeting, the ISS partners agreed on a process for selecting a final ISS configuration by December 2003. The 2003 space shuttle *Columbia* tragedy delayed the process, and President Bush's January 2004 announcement of the Vision for Space Exploration, changed NASA's own plans, including cancellation of the Orbital Space Plane, and termination of the space shuttle program after ISS construction is completed. At a January 2005 Heads of Agency meeting, the partners endorsed a final configuration of ISS, but NASA subsequently announced changes to it. The agency now plans to conduct only 18 (instead of 28) launches to the ISS, all before the end of 2010, and has dropped plans to launch the centrifuge and its accommodation module, and Russia's Science Power Platform. The agency plans to meet with the other ISS partners to discuss these changes in late 2005.

The changes to the ISS are largely due to the new direction NASA is taking in response to the Vision for Space Exploration. Inter alia, the Vision calls for development of a Crew Exploration Vehicle (CEV) to take astronauts to and from the Moon. It also can take them to and from the ISS, and NASA Administrator Griffin stated at a September 19, 2005 press conference that the CEV would be used to take crews to and from the ISS, and to serve as a lifeboat for them. If the CEV is built as announced, it would fulfill the U.S. commitment to build a crew return capability, and allow the ISS crew size to increase to its originally planned complement of seven. President Bush directed that the CEV be ready by 2014; Dr. Griffin hopes to accelerate the schedule to 2012.

Congressional Action

FY2005

The FY2005 request for the ISS program was $2.412 billion: $1.863 billion for construction and operations, including $140 million in a new "ISS Crew/Cargo Services" line to pay for alternatives to the shuttle for taking crew and cargo to and from ISS; and $549 million for research. Congress did not specify a funding level for the ISS in the final version of the FY2005 VA-HUD-IA appropriations act, which was incorporated in the FY2005 Consolidated Appropriations Act (H.R. 4818, P.L. 108-447). Instead, it gave NASA "unrestrained transfer authority" to shift money between budget accounts. In a May 10, 2005 update to its FY2005 operating plan, NASA indicated that it is shifting $160 million from the space station into the space shuttle program for costs associated with returning the shuttle to flight status. The operating plan shows $1.676 billion for ISS construction and operations, including $98 million for ISS Crew/Cargo Services. However, management of the ISS Crew/Cargo Services activity, and the $98 million, have been moved to the Exploration Systems Mission Directorate. The operating plan retains the $98 million in the ISS subaccount, however. The FY2005 total shown in the table in this issue brief — $2.058 billion — is the sum of $1.676 billion as shown in NASA's May 2005 operating plan, plus $382 million allocated for space station research as shown in a FY2006 NASA budget chart.

FY2006

For FY2006, NASA originally requested $2.180 billion for the ISS program: $1.857 billion for construction and operations (including $160 million for ISS Crew/Cargo Services), and $324 million for ISS research. In a May 10, 2005 FY2005 operating plan update, NASA announced that it was moving the ISS Crew/Cargo Services activity to the Exploration Systems Mission Directorate (ESMD). A July 15 amended budget request also moved the FY2006 funding for that activity (which it said was $168 million) into ESMD and commensurately reduced the ISS construction and operations request. NASA later also reduced the amount it plans to spend on ISS research to $306 million. The number used in the accompanying table ($1,995 million) reflects these changes.

Table 1. U.S. Space Station Funding (in $ millions)

Fiscal Year	Request	Appropriated
1985	150	150
1986	230	205
1987	410	410
1988	767	425
1989	967	900
1990	2,050	1,750
1991	2,430	1,900
1992	2,029	2,029
1993	2,250	2,100
1994	2,106	2,106
1995	2,113	2,113
1996	2,115	2,144
1997	2,149	2,149
1998	2,121	2,441A
1999	2,270	2,270
2000	2,483	2,323
2001	2,115	2, 115
2002	2,114	2,093
2003	1,839	1,810 B
2004 C	2,285	2,085
2005	2,412	2,058 D
2006	1,995	

The numbers here reflect NASA's figures for "the space station program." Over the years, what is included in that definition has changed. The appropriated amount may differ from actual spending.

[a] NASA's FY1999 budget documents showed $2.501 billion in the expectation Congress would approve additional transfer requests, but it did not.

[b] Adjusted for 0.65% rescission.

[c] Reflects shift to full cost accounting.

[d] Congress did not specify an appropriations level. This figure is from NASA's May 10, 2005 operating plan, plus the amount to be spent on ISS research.

The FY2006 Science, State, Justice and Commerce (SSJC) appropriations bill (H.R. 2862), includes NASA. The House cut $10 million from ISS construction and operations, $10 million from ISS Crew/Cargo Services, and $25 million from the account that funds research on ISS, though it did not specify that the cuts come from the ISS portion of that account. The Senate cut all $160 million from

ISS Crew/Cargo Services. The final version of the bill cut $80 million, including $60 million from ISS Crew/Cargo Services (the conference report does not reflect the July budget amendment.) S. 1281, a FY2006-2010 NASA authorization bill, adds $100 million in FY2006 and makes other changes to enhance ISS research. H.R. 3070 (H.Rept. 109-173) is a FY2006-2007 NASA authorization bill, that does not specify an amount for ISS.

INTERNATIONAL PARTNERS

The Original Partners: Europe, Canada, and Japan

Canada, Japan, and most of the 16 members of the European Space Agency (ESA) have been participating in the space station program since it began. Formal agreements were signed in 1988, but had to be revised following Russia's entry into the program, and two more European countries joined in the interim. The revised agreements were signed on January 29, 1998, among the partners in the ISS program: United States, Russia, Japan, Canada, and 11 European countries — Belgium, Denmark, France, Germany, Italy, the Netherlands, Norway, Spain, Sweden, Switzerland, and the United Kingdom. Representatives of the various governments signed the government-to-government level Intergovernmental Agreement (IGA) that governs the program. (The United Kingdom signed the IGA, but is not financially participating in the program so the number of European countries participating in the program is variously listed as 10 or 11.) NASA also signed Memoranda of Understanding for implementing the program with its counterpart agencies: the European Space Agency (ESA), the Canadian Space Agency (CSA), the Russian space agency (then Rosaviakosmos, now Roskosmos), and the Japanese Science and Technology Agency. The IGA is a treaty in all the countries except the United States (where it is an Executive Agreement).

Canada is contributing the Mobile Servicing System (MSS) for assembling and maintaining the space station. In February 1994, the new prime minister of Canada had decided to terminate Canada's role in the program, but later agreed to reformulate Canada's participation instead. The first part of the MSS (the "arm" or Canadarm 2) was launched in April 2001; another part, the Special Purpose Dextrous Manipulator, is awaiting launch.

ESA is building a laboratory module called Columbus, and an Automated Transfer Vehicle (ATV) to take cargo to ISS. The ATV will be launched on Europe's Ariane launch vehicle. The first ATV launch is expected in 2006. The major contributors to Columbus are Germany, France, and Italy. Budgetary difficulties over the years led ESA to cancel other hardware it was planning. ESA also is building a cupola (a windowed dome) and two of the three "nodes" (Node 2 and Node 3) for NASA in exchange for NASA launches of Europe's module and other services. Node 2, Columbus, and the cupola are awaiting launch. NASA had canceled plans for Node 3, but now has revived them. NASA also has a bilateral agreement with Italy under which Italy built three "mini-pressurized logistics modules" (MPLMs). Already in use, they are launched via the shuttle, attached to ISS while cargo is transferred to the station, filled with refuse or other unwanted material, placed back into the shuttle's cargo bay, and returned to Earth.

Japan is building the Japanese Experiment Module, named Kibo (Hope). One part is pressurized and the other is not (called the "back porch," it will be exposed to space for experiments requiring those conditions). The pressurized section is awaiting launch; the unpressurized section is in development. Japan also is building a centrifuge and a Centrifuge Accommodation Module ("CAM") for NASA in exchange for shuttle flights to launch Kibo, but NASA decided in September 2005 that it no longer needs the centrifuge. NASA plans to meet with Japan and the other partners in late 2005 to discuss this and other matters.

CSA reported in February 2004 that Canada's total ISS spending is expected to be $1.4 billion (Canadian), of which $1.3 billion (Canadian) was spent by that time. ESA reported in March 2004 that its estimated funding for ISS is 5.1 billion Euros, of which 4.1 billion Euros were spent as of the end of 2003. (In March 2004, 1 Canadian dollar = 0.75 U.S. dollar and 1 Euro = 1.2 U.S. dollars.) In February 2004, the Japanese space agency reported that Japan expects to spend $4.8 billion on ISS, of which $4 billion was spent by the end of March 2003. (A bilateral agreement was signed with Brazil in October 1997 for Brazil to provide payload and logistics hardware. Brazil is restructuring its agreement in light of financial constraints, however. The level of its funding contribution is unclear.)

Russia

Issues associated with Russia's participation in ISS are discussed elsewhere. This section explains Russian space station activities from 1971 to the present. The Soviet Union launched the world's first space station, Salyut 1, in 1971 followed by five more *Salyuts* and then *Mir*. At least two other *Salyuts* failed before they could be occupied. The Soviets accumulated a great deal of data from the many missions flown to these stations on human adaptation to weightlessness. The data were often shared with NASA. They also performed microgravity materials processing research, and astronomical and Earth remote sensing observations. Importantly, they gained considerable experience in operating space stations. Russia's most recent space station, *Mir*, was a modular space station built and operated between 1986 and 2001. Crews were ferried back and forth to *Mir* using Soyuz spacecraft.

Crews occupied *Mir* from 1986-2000. For almost ten of those years (1989-1999), *Mir* was continuously occupied by crews on a rotating basis. Although occasionally crews stayed for very long periods of time to study human reaction to long duration spaceflight, typically they remained for 5-6 months and then were replaced by a new crew. From 1995-1998, seven Americans participated in long duration (up to six month) missions aboard *Mir*, and nine space shuttle missions docked with the space station. Individuals from Japan, Britain, Austria, Germany, France, and the Slovak Republic also paid for visits to *Mir*. Russia deorbited *Mir* into the Pacific Ocean on March 23, 2001.

Risks and Benefits of Russian Participation

For many years, controversy over the ISS program focused on Russia's participation in the program. Among the issues were the extent to which successful completion of ISS is dependent on Russia, Russia's financial ability to meet its commitments, and whether the United States should provide funding to Russia if it proliferates missile technology to certain countries. While there is no exchange of funds among the other ISS partners, the United States (and other partners) have provided funding to Russia. By 1998, the United States had paid approximately $800 million to Russia for space station cooperation.

Following the Clinton Administration's decision to bring Russia into the program, Congress stated that Russian participation "should enhance and not enable" the space station (H.Rept. 103-273, to accompany H.R. 2491, the FY1994 VA-HUD-IA appropriations act, P.L. 103-124). The current design, however, can only be viewed as being "enabled" by Russian participation. It is dependent on Russian Progress vehicles for reboost (to keep the station from reentering Earth's atmosphere), on Russian Soyuz spacecraft for emergency crew return, and on Russia's Zvezda module for crew quarters (which allows ISS to be permanently occupied). When the shuttle is unavailable, U.S. access to ISS is completely dependent on Russia, which ferries crews back and forth on the Soyuz spacecraft and takes cargo to ISS on Progress spacecraft. President Bush's exploration initiative would increase U.S. dependence on Russia vis a vis the space station (see Key Issues for Congress below).

Russia's financial ability to meet its commitments is an ongoing issue. The launch of Zvezda, the first module Russia had to pay for itself, was more than two years late. (Zarya was built by Russia, but NASA paid for it.) Since Zvezda's launch in 2000, Russia has met its commitments to launch Soyuz and Progress spacecraft, but is reassessing what other modules and hardware it will build at its own expense. Russian space agency officials have repeatedly expressed concern about whether they can provide the needed number of Soyuz and Progress spacecraft because of budget constraints.

Clinton Administration and NASA officials asserted repeatedly that Russian participation in the space station program would accelerate the schedule by two years and reduce U.S. costs by $4 billion. That was later modified to one year and $2 billion, and an April 1, 1994 letter to Congress from NASA said 15 months and $1.5 billion. NASA officials continued to use the $2 billion figure thereafter, however. GAO concluded (GAO/NSIAD 94-248) that Russian participation would cost NASA $1.8 billion, essentially negating the $2 billion in expected savings. In 1998, a NASA official conceded that having Russia as a partner added $1 billion to the cost. Other benefits cited by the Clinton Administration were providing U.S. financial assistance to Russia as it moves to a market economy, keeping Russian aerospace workers employed in non-threatening activities, and the emotional impact, historic symbolism, and potential long term significance of the two former Cold War adversaries working together in space.

One benefit is that the space station can be serviced with Russian as well as American spacecraft, providing redundancy if either side must ground its fleet due to an accident, for example. This is an important advantage while the U.S. space shuttle is grounded. Russia is providing both crew and cargo flights to the space station, enabling it to continue operation without the shuttle.

ISS and U.S. Nonproliferation Objectives, Including the Iran Nonproliferation Act (INA)

The overall relationship between the United States and Russia is another factor in the ISS equation, including Russian adherence to U.S. nonproliferation objectives. Getting Russia to adhere to the Missile Technology Control Regime (MTCR), designed to stem proliferation of ballistic missile technology, appears to have been a primary motivation behind the Clinton Administration's decision to add Russia as a partner. The United States wanted Russia to restructure a contract with India that would have given India advanced rocket engines and associated technology and know-how. The United States did not object to giving India the engines, but to the technology and know-how. Russia claimed that restructuring the contract would cost $400 million. The 1993 agreement to bring Russia into the space station program included the United States paying Russia $400 million for space station cooperation. At the same time, Russia agreed to adhere to the MTCR. The question is what the United States will do if Russia violates the MTCR. Some Members of Congress believe Russia already has done so. The Clinton Administration sanctioned 10 Russian entities for providing technology to Iran. Neither the Russian space agency nor any major Russian ISS contractors or subcontractors were among those sanctioned.

On March 14, 2000, President Clinton signed into law the Iran Nonproliferation Act (INA), P.L. 106-178. The law, *inter alia*, prohibits NASA from making "extraordinary payments" related to ISS after January 1, 1999, in cash or in kind, to Russia unless Russia takes the necessary steps to prevent the transfer of weapons of mass destruction and missile systems to Iran, and the President makes a determination that neither the Russian space agency nor any entity reporting to it has made such transfers for at least one year prior to such determination. Exceptions are made for payments needed to prevent imminent loss of life by or grievous injury to individuals aboard ISS (the "crew safety" exception); for payments to construct, test, prepare, deliver, launch, or maintain Zvezda as long as the funds do not go to an entity that may have proliferated to Iran and the United States receives goods or services of commensurate value; and hardware needed to dock the U.S. Interim Control Module (ICM, discussed earlier). Certain notifications are required if the exceptions are utilized. NASA was seeking permission to spend $35 million on Russia goods and services, of which $14 million was for the ICM docking hardware. President Clinton provided Congress with a required notification with regard to that $14 million on June 29, 2000. Ultimately, only $11 million was needed for the ICM hardware, leaving $24 million that NASA wanted to spend. No determination as required by the act was forthcoming from the President. NASA considered using the crew safety exception, but at a House International Relations Committee hearing on October 12, 2000, some committee Members sharply criticized NASA's legal interpretation of that exception, particularly NASA's broad interpretation of the word "imminent."

Thus, the INA has important ramifications for whether NASA can keep its astronauts on ISS for long duration missions after April 2006, or at all after 2010 if the shuttle is terminated as planned (see Key Issues for Congress, below). The Bush Administration submitted a proposed amendment to Congress on July 12, 2005, to modify the INA. The amendment would have prohibited payments only for goods or services that Russia had previously agreed to provide at no cost. On

September 21, 2005, the Senate passed S. 1713 (Lugar) which would have allowed NASA to pay Russia, but only through January 1, 2012. The House passed an amended version on October 26 that, inter alia, clarifies that the goods or services must be delivered, as well as paid for, by January 1, 2012. (The House version makes other changes as well that are outside the scope of this report to discuss.) The Senate agreed to the House-passed version on November 8. The bill is awaiting signature by the President. See CRS Report RS22270 for more information.

KEY ISSUES FOR CONGRESS

Maintaining ISS Operations while the Shuttle Is Grounded

The grounding of the space shuttle system following the *Columbia* accident suspended assembly of ISS, temporarily reduced the size of Expedition crews from three to two, and complicated efforts to keep the crews supplied with consumables, scientific experiments, and spare parts for equipment that needs repair. NASA launched the shuttle on its first Return to Flight mission, STS-114, on July 26, 2005 (see CRS Report RS21408). STS-114 landed successfully on August 9, after spending much of its mission docked with ISS. Another shuttle launch was scheduled for September, but a problem occurred during STS-114's launch and NASA has again grounded the fleet. The next shuttle mission is not expected until at least May 2006.

In the absence of the shuttle, ISS crews are taken to and from ISS using Russian Soyuz spacecraft on the same six-month schedule already planned, and Russian Progress spacecraft resupply the crew. Russia is obligated to provide crew return for three people throughout the lifetime of ISS. Currently, they accomplish that with two Soyuzes per year (each lasts only six months once docked to ISS). Russia also is obligated to provide a certain number of Progress spacecraft, but has cautioned that funding for Soyuz and Progress is not assured.

The Russians operated seven of their own space stations (see above) using only Soyuz and Progress spacecraft, so it is possible to keep ISS operating without the shuttle as long as Russia is willing to provide them. However, operation of ISS was premised on the availability of the cargo-carrying capacity of the space shuttle. Keeping ISS operating without the shuttle is challenging. For example, the Expedition 10 crew was required to reduce its food intake because of shortages aboard the station in late 2004. Stocks were resupplied by a Progress that reached ISS late that December, but U.S. and Russian space station personnel made clear that if the Progress had failed to dock, the crew would have had to return home prematurely because of the food situation. ISS crews also need to repair faulty equipment, but replacement parts may not fit aboard Progress or Soyuz. NASA's decision to again ground the shuttle fleet may lead to questions about whether to keep a crew aboard ISS. In addition to questions about keeping the crews well supplied, with a two-person crew, less time may be available for scientific research, and without the shuttle, fewer experiments can be taken to ISS. If little scientific research can be accomplished, some may question the wisdom of asking astronauts and cosmonauts to accept the risks inherent in human spaceflight simply to maintain ISS systems. Conversely, how long ISS could continue to function with no one aboard is unknown. Progress spacecraft could dock with ISS automatically to reboost it and keep it at the proper altitude, but a major system malfunction that could not be remedied by ground-based controllers could imperil the station.

Ensuring U.S. Astronaut Participation in Long-Duration Missions

As noted, ISS crew size is limited in part by the number of occupants that can be accommodated in a lifeboat, or "crew return vehicle" (CRV), in an emergency. Without its own CRV until at least 2012, NASA faced two deadlines in terms of assuring access to ISS for long duration missions by U.S. astronauts. First was spring 2006, when Russia will have fulfilled its commitment to provide lifeboat services for U.S. astronauts. Second was 2010, when the shuttle is expected to be retired in accordance with President Bush's Vision for Space Exploration. The 2010 deadline is discussed separately (see next issue).

Regarding the spring 2006 deadline, the international agreements that govern the ISS program obligate Russia to provide crew return services for three crew members throughout the lifetime of the ISS, and the United States to provide such services for at least four people beginning when assembly of ISS is completed. Prior to the *Columbia* tragedy, that milestone was expected to occur in 2006. By that time, a U.S. CRV, accommodating at least four people, was expected to be available. It would not only provide lifeboat capabilities for U.S. crew members, but allow the ISS crew size to grow to seven (three could return on a Soyuz, and four on the CRV), including representatives from the other partners. The Bush Administration terminated the CRV, and its successor, the Orbital Space Plane. NASA now has resumed planning to build a spacecraft that could serve this function (the Crew Exploration Vehicle), but it is not expected to be ready until at least 2012. Therefore, only Russian Soyuz spacecraft are currently available for lifeboat services.

Under a 1996 "balance agreement" between NASA and the Russian space agency, Russia is obligated to use 11 Soyuz spacecraft to provide crew return for U.S. crews. The last of the 11 was launched on October 1, 2005, and is expected to return to Earth in April 2006. After that, Russia no longer must allocate any of the seats on its Soyuzes for U.S. astronauts. It can sell those flight opportunities to whomever it wishes. Russia is interested in selling flight opportunities to help finance its space program, so might choose to limit those opportunities to paying customers. As noted, the INA prevents NASA from paying Russia for ISS-related activities unless Russia does not proliferate certain technologies to Iran. Congress has passed legislation (S. 1713) that allows NASA to purchase ISS-related goods and services through January 1, 2012. The bill is awaiting signature by the President.

Impact of President Bush's Vision for Space Exploration, Including a Potential Gap in U.S. Human Access to Space

President Bush's January 2004 Vision for Space Exploration directs NASA to focus its activities on returning humans to the Moon by 2020, and someday sending them to Mars and "world beyond." The Vision affects the ISS program in several ways. First, the President directed that the shuttle be retired in 2010, and NASA officials indicated that they would complete their use of the ISS in 2016. By terminating the shuttle and NASA utilization of the ISS, that funding (approximately $6 billion per year) could be reallocated to achieving other aspects of the Vision. The President also directed that the U.S. research on ISS be restricted to research needed to support the Vision, instead of the broadly-based research program that was planned.

The extent to which ISS can be utilized without the space shuttle is not clear. Soyuz spacecraft can take crews back and forth, but the shuttle's cargo capacity — both for taking cargo to ISS, and back to Earth (e.g. the results of scientific experiments, or hardware that needs repair) — could be expensive to replicate. No other partner has a spacecraft able to bring material back to Earth today. Europe reportedly is considering adding a return capsule to its Automated Transfer Vehicle (a robotic cargo spacecraft that is expected to make its first flight in 2007). NASA is exploring the possibility of U.S. commercial companies providing launch and/or return services through the ISS Crew/Cargo activity.

Another issue was how U.S. crews could access ISS once the shuttle is retired in 2010. As already discussed, NASA was facing an April 2006 deadline regarding Russian support for U.S. astronauts because of the Iran Nonproliferation Act (INA). That issue appears to be resolved with congressional passage of S. 1713, though that allows NASA to purchase Russian goods and services only through January 1, 2012. If NASA's new CEV is not available by that time, additional modifications to the INA may be required. Once the shuttle is retired in 2010, NASA will not be able to launch its own astronauts into space at all until the CEV is ready.

LEGISLATION

H.R. 2862 (Wolf). FY2006 Science, State, Justice, Commerce Appropriations Act (includes NASA). Reported from House Appropriations Committee June 10, 2005 (H.Rept. 109-118); passed House June 16. Reported from Senate Appropriations Committee June 23 (S.Rept. 109-88); passed Senate September 15, 2005. Conference report (H.Rept. 109-272) filed November 7, passed House November 9, passed Senate November 16.

H.R. 3070 (Calvert)/S. 1281 (Hutchison). NASA authorization bill. H.R. 3070 authorizes NASA funding for FY2006-2007; reported by the House Science Committee July 18 (H.Rept. 109-173), passed House, amended, July 22, 2005. S. 1281 authorizes NASA funding for FY2006-2010; reported by Senate Commerce Committee July 26 (S.Rept. 109-108), passed Senate, amended, September 28, 2005.

S. 1713 (Lugar). Amends the Iran Nonproliferation Act to allow NASA to purchase ISS-related services from Russia through January 1, 2012. Discharged from Senate Foreign Relations Committee and passed Senate September 21. Passed House, amended, October 26, 2005. Senate agreed with House-amended version, November 8. Presented to the President November 17.

In: Space Policy and Exploration
Editor: W. N. Callmers, pp. 87-99
ISBN: 978-1-60456-448-8

Chapter 4

THE INTERNATIONAL SPACE STATION AND THE SPACE SHUTTLE*

Carl E. Behrens

ABSTRACT

The International Space Station (ISS) program began in 1993, with Russia joining the United States, Europe, Japan, and Canada. Crews have occupied ISS on a 4-6 month rotating basis since November 2000.

The U.S. Space Shuttle, which first flew in April 1981, has been the major vehicle taking crews and cargo back and forth to ISS, but the shuttle system has encountered difficulties since the *Columbia* disaster in 2003. Russian Soyuz spacecraft are also used to take crews to and from ISS, and Russian Progress spacecraft deliver cargo, but cannot return anything to Earth, since they are not designed to survive reentry into the Earth's atmosphere. A Soyuz is always attached to the station as a lifeboat in case of an emergency.

President Bush, prompted in part by the *Columbia* tragedy, made a major space policy address on January 14, 2004, directing NASA to focus its activities on returning humans to the Moon and someday sending them to Mars. Included in this "Vision for Space Exploration" is a plan to retire the space shuttle in 2010. The President said the United States would fulfill its commitments to its space station partners, but the details of how to accomplish that without the shuttle were not announced.

The shuttle *Discovery* was launched on July 4, 2006, and returned safely to Earth on July 17. This was the first of 16 *post-Columbia* flights to the ISS

* Excerpted from CRS Report RL33568, dated April 26, 2007.

that NASA plans to complete before retiring the shuttle in 2010. The shuttle *Atlantis* followed with a September 9 launch that resumed construction of the International Space Station before returning to earth September 21. *Discovery* returned to the ISS in December, returning to earth December 22 after a 13-day mission that included four space walks and continued work on the station.

On June 29, 2006, the House passed the FY2007 Science, Justice, Commerce and Related Agencies appropriations bill (H.R. 5672), which includes funding for NASA. The bill would have funded the Exploration Capabilities account, which includes the shuttle and the ISS, at $6.194 billion in FY2007. The Senate Appropriations Committee reported its version of H.R. 5672 (S.Rept. 109-280) on July 13, 2006, recommending $6.235 billion for Exploration Capabilities, but the Senate did not take action on the bill. NASA programs were funded for FY2007 in the Revised Continuing Appropriations Resolution, 2007 (H.J.Res. 20, P.L. 110-5). P.L. 110-5 funded Exploration Capabilities at $6.140 billion.

MOST RECENT DEVELOPMENTS

The space shuttle *Atlantis,* scheduled to launch in March 2007 but damaged in a hailstorm in late February, was expected to be repaired and ready for launch in June or July. The mission, STS-117, is aimed at continuing construction of the International Space Station (ISS).

NASA will schedule a shuttle flight in May 2008 to service the Hubble Space Telescope, Administrator Michael Griffin announced on October 31, 2006. The mission would extend the life of the telescope through 2013. The decision reverses a 2004 cancellation of the servicing mission following the loss of the shuttle *Columbia.*

National Aeronautics and Space Administration (NASA) programs were funded for FY2007 in the Revised Continuing Appropriations Resolution, 2007 (H.J.Res. 20, P.L. 110-5). The Exploration Capabilities program, which includes the space shuttle and the ISS, were funded at $6.140 billion, compared to $6.516 billion in FY2006. The request for FY2008 for Exploration Capabilities was $6.792 billion.

THE INTERNATIONAL SPACE STATION (ISS)

NASA launched its first space station, Skylab, in 1973. Three crews were sent to live and work there in 1973-1974. It remained in orbit, unoccupied, until it reentered Earth's atmosphere in July 1979, disintegrating over Australia and the Indian Ocean. Skylab was never intended to be permanently occupied, but the goal of a permanently occupied space station with crews rotating on a regular basis, employing a reusable space transportation system (the space shuttle) was high on NASA's list for the post-Apollo years following the moon landings. Budget constraints forced NASA to choose to build the space shuttle first. The first launch of the shuttle was in April 1981. When NASA declared the shuttle "operational" in 1982, it was ready to initiate the space station program.

In his January 25, 1984 State of the Union address, President Reagan directed NASA to develop a permanently occupied space station within a decade, and to invite other countries to join. On July 20, 1989, the 20th anniversary of the first Apollo landing on the Moon, President George H. W. Bush voiced his support for the space station as the cornerstone of a long-range civilian space program eventually leading to bases on the Moon and Mars. That "Moon/Mars" program, the Space Exploration Initiative, was not greeted with enthusiasm in Congress, primarily due to budget concerns, and ended in FY1993, although the space station program continued.

President Clinton dramatically changed the character of the space station program in 1993 by adding Russia as a partner to this already international endeavor. That decision made the space station part of the U.S. foreign policy agenda to encourage Russia to abide by agreements to stop the proliferation of ballistic missile technology, and to support Russia economically and politically as it transitioned from the Soviet era. The Clinton Administration strongly supported the space station within certain budget limits.

The International Space Station program thus began in 1993, with Russia joining the United States, Europe, Japan, and Canada. An Intergovernmental Agreement (IGA) established three phases of space station cooperation. The IGA is a treaty in all the countries except the United States, where it is an Executive Agreement. It is implemented through Memoranda of Understanding (MOUs) between NASA and its counterpart agencies.

During Phase I (1995-1998), seven U.S. astronauts remained on Russia's space station *Mir* for long duration (several month) missions with Russian cosmonauts, Russian cosmonauts flew on the U.S. space shuttle seven times, and nine space shuttle missions docked with *Mir* to exchange crews and deliver supplies. Repeated system failures and two life-threatening emergencies on *Mir* in

1997 raised questions about whether NASA should leave more astronauts on *Mir*, but NASA decided *Mir* was sufficiently safe to continue the program. (*Mir* was deorbited in 2001.) Phases II and III involve construction of the International Space Station itself, and blend into each other. Phase II began in 1998 and was completed in July 2001; Phase III is underway.

President George W. Bush, prompted in part by the February 2003 space shuttle *Columbia* tragedy, made a major space policy address on January 14, 2004, directing NASA to focus its activities on returning humans to the Moon and eventually sending them to Mars. Included in this "Vision for Space Exploration" is a plan to retire the space shuttle in 2010. The President said the United States would fulfill its commitments to its space station partners, but the details of how to accomplish that without the shuttle were not announced.

ISS Design, Cost, Schedule, and Lifetime

Under the original ISS schedule, assembly of the station would have been completed in 2002, with operations at least through 2012. President Bush restructured the space station program in 2001, and left it unclear when assembly would be completed. NASA briefing charts in March 2003 showed space station operations possibly continuing until 2022. Under President Bush's January 2004 "Vision for Space Exploration," however, NASA plans to complete its utilization of ISS in 2016 (though the other partners may continue to use it after that time).

ISS segments are launched into space on U.S. or Russian launch vehicles and assembled in orbit. The space station is composed of a multitude of modules, solar arrays to generate electricity, remote manipulator systems, and other elements. (Details can be found at [http://spaceflight.nasa.gov/home/index.html].) Six major modules are now in orbit. The first two were launched in 1998: Zarya ("Sunrise," a Russian-built, U.S.-owned, module with guidance, navigation, and control systems) and Unity (a U.S. "node" connecting other modules). Next was Zvezda ("Star," a Russian module that serves as the crew's living quarters) in 2000. Destiny (a U.S. laboratory), Quest (a U.S. airlock), and Pirs ("Pier," a Russian docking compartment) arrived in 2001. Among the other modules awaiting launch are laboratory modules built by Russia, Europe, and Japan, and two more "nodes" built by Europe. (Zarya counts as a U.S. module because NASA paid Russia to build it. The European-built nodes and Cupola count as U.S. components because they were built under barter agreements where Europe produces hardware NASA needs instead of paying cash to NASA for launch and other ISS-related services.

Japan was to build a centrifuge and its Centrifuge Accommodation Module under such a barter arrangement, but NASA terminated that activity in 2005.)

The U.S. space shuttle has been the major vehicle taking crews and cargo back and forth to ISS, but the shuttle system has encountered difficulties since the *Columbia* disaster. Russian Soyuz spacecraft are also used to take crews to and from ISS, and Russian Progress spacecraft deliver cargo, but cannot return anything to Earth, since it is not designed to survive reentry into the Earth's atmosphere. A Soyuz is always attached to the station as a lifeboat in case of an emergency.

"Expedition" crews have occupied ISS on a 4-6 month rotating basis since November 2000. Originally the crews had three members (two Russians and one American, or two Americans and one Russian), with an expectation that crew size would grow to six or seven once assembly was completed. Crew size was temporarily reduced to two (one American, one Russian) while the U.S. shuttle was grounded in order to reduce resupply requirements. The number of astronauts who can live on the space station is limited in part by how many can be returned to Earth in an emergency by lifeboats docked to the station. Only Russian Soyuz spacecraft are available as lifeboats. Each Soyuz can hold three people, limiting crew size to three if only one Soyuz is attached. NASA planned to build a U.S. Crew Return Vehicle (CRV) to provide lifeboat capabilities for at least four more crew. The Bush Administration canceled those plans due to cost growth in the ISS program, then began a different program (the Orbital Space Plane) that also was cancelled. In September 2005, NASA announced that the new Crew Exploration Vehicle (CEV) it is building to implement the President's Vision for Space Exploration (the "Moon/Mars" program) will be designed to take crews to and from the ISS, and to serve as a lifeboat. NASA currently hopes to have it ready by 2012.

Each Soyuz must be replaced every six months. The replacement missions are called "taxi" flights since the crews bring a new Soyuz up to ISS and bring the old one back to Earth. Therefore, under normal conditions, the long duration Expedition crews are regularly visited by taxi crews, and by the space shuttle bringing up additional ISS segments or exchanging Expedition crews. When the shuttle is unavailable, Expedition crews are taken back and forth on the "taxi" flights.

Space Station Costs

From FY1994-FY2001, the cost estimate for building ISS grew from $17.4 billion to about $25 billion. The $17.4 billion estimate did not include launch costs, operational costs after completion of assembly, civil service costs, or other costs. NASA estimated the program's life-cycle cost (all costs, including funding spent prior to 1993) from FY1985-FY2012 at $72.3 billion. In 1998, GAO estimated the life-cycle cost at $95.6 billion (GAO/NSIAD-98-147). More recent, comparable, life-cycle estimates are not available from NASA or GAO.

As costs continued to rise, Congress voted to legislate a $25 billion cap on development of the ISS program, plus $17.7 billion for associated shuttle launches, in the FY2000-FY2002 NASA authorization act (P.L. 106-391). In January 2001, however, NASA announced that the cost would be over $30 billion, 72% above the 1993 estimate, and $5 billion above the legislated cap. NASA explained that program managers had underestimated the complexity of building and operating the station. The Bush Administration signaled it supported the legislated cap, would not provide additional funds, and NASA would have to find what it needed from within its Human Space Flight account.

"Core Complete" Configuration

In February 2001, the Bush Administration announced it would cancel or defer some ISS hardware to stay within the cap and control space station costs. The decision truncated construction of the space station at a stage the Administration called "core complete." In 2001, the space station program office at Johnson Space Center (JSC) estimated that it would cost $8.3 billion from FY2002-FY2006 to build the core complete configuration, described at that time as all the U.S. hardware planned for launch through "Node 2," plus the launch of laboratories being built by Europe and Japan. NASA subsequently began distinguishing between "U.S. Core Complete" (the launches through Node 2, which, prior to the *Columbia* tragedy, was scheduled for February 2004) and "International Partner (IP) Core Complete" which included the addition of European and Japanese laboratory modules (then anticipated in 2008).

The new policy was followed by President Bush's January 2004 "Vision for Space Exploration," which directs that U.S. research on ISS be restricted only to that which supports the Vision. A new research plan, incorporating the President's Vision, has not been released by NASA. However, the 2005 NASA authorization act (P.L. 109-155), directs that at least 15% of ISS research spending be non-Vision-related (Sec. 204).

At a January 2005 Heads of Agency meeting, the partners endorsed a final configuration of ISS, but NASA subsequently announced changes to it. The agency now plans to conduct only 16 (instead of 28) shuttle launches to the ISS, all before the end of FY2010 (September 30, 2010), and has dropped plans to launch the centrifuge and its accommodation module, and Russia's Science Power

Platform. The agency plans to meet with the other ISS partners to discuss these changes.

Table 1. U.S. Space Station Funding (in $ millions)

Fiscal Year	Request	Appropriated
1985	150	150
1986	230	205
1987	410	410
1988	767	425
1989	967	900
1990	2,050	1,750
1991	2,430	1,900
1992	2,029	2,029
1993	2,250	2,100
1994	2,106	2,106
1995	2,113	2,113
1996	2,115	2,144
1997	2,149	2,149
1998	2,121	2,441 [a]
1999	2,270	2,270
2000	2,483	2,323
2001	2,115	2,115
2002	2,114	2,093
2003	1,839	1,810
2004 [b]	2,285	2,085
2005	2,412	2,058
2006	1,995	1,972
2007	1,894	NA[c]
2008	1,894	2,239

These numbers reflect NASA's figures for "the space station program." Over the years, what is included in that definition has changed. In recent years, funding for ISS research has been located in a different account from ISS development funding. The figures here represent the ISS development and ISS research request and appropriations to the maximum extent possible.

a. NASA's FY1999 budget documents showed $2.501 billion in the expectation Congress would approve additional transfer requests, but it did not.

b. Reflects shift to full cost accounting.

c. Space shuttle funding was not specified in P.L. 110-5.

The changes to the ISS are largely due to the new direction NASA is taking in response to the Vision for Space Exploration. The Vision calls for development of a Crew Exploration Vehicle, now named Orion, to take astronauts to and from the Moon, and a Crew Launch Vehicle, now named Ares I. Orion also can take them to and from the ISS, and NASA Administrator Griffin stated at a September 19, 2005 press conference that Orion would be used to take crews to and from the ISS, and to serve as a lifeboat for them. If Orion is built as announced, it would fulfill the U.S. commitment to build a crew return capability, and allow the ISS crew size to increase to its originally planned complement of seven. An Earth-orbit capability is planned by 2014 (although NASA now considers early 2015 more likely) with the ability to take astronauts to and from the Moon following no later than 2020.

THE SPACE SHUTTLE

The Space Transportation System (STS) — the Space Shuttle — is a partially reusable launch vehicle and is the sole U.S. means for launching humans into orbit. It consists of an airplane-like Orbiter, with two Solid Rocket Boosters (SRBs) on each side, and a large, cylindrical External Tank (ET) that carries fuel for the Orbiter's main engines. The Orbiters and SRBs are reused; the ET is not. NASA has three remaining spaceflight-worthy Orbiters: *Discovery*, *Atlantis*, and *Endeavour*.

The *Challenger* and *Columbia* Tragedies

A total of 117 shuttle launches have taken place since April 1981. Two ended in tragedy, each killing seven astronauts. In 1986, the space shuttle *Challenger* exploded 73 seconds after launch because of the failure of a seal (an O-ring) between two segments of an SRB. In 2003, the space shuttle *Columbia* disintegrated as it returned to Earth after 16 days in orbit (see CRS Report RS21408, *NASA's Space Shuttle Program: The Columbia Tragedy, the Discovery Mission, and the Future of the Shuttle*, by Marcia S. Smith). A hole in *Columbia's* left wing, caused during launch by a piece of foam insulation that detached from the ET, allowed hot gases to enter the wing during reentry, deforming it and causing the shuttle to break up. The *Columbia* Accident Investigation Board (CAIB) found that the tragedy was caused by technical and organizational failures, and made 29 recommendations, 15 of which it said should be completed before the shuttle returned to flight.[1]

Sean O'Keefe, NASA's Administrator from December 2001-February 2005, said NASA would comply with the CAIB recommendations. He established an RTF ("Return to Flight") Task Group, chaired by two former astronauts, Tom

Stafford and Dick Covey, to oversee NASA's implementation of the CAIB's 15 RTF recommendations. The Stafford/Covey Task Group [http://www.return toflight.org] did not address management and culture changes, and was not tasked to determine whether the shuttle was ready to return to flight. Its assignment only was to evaluate NASA's compliance with the CAIB recommendations for RTF. The Task Group ultimately concluded that NASA met the intent of 12 of the 15 CAIB RTF recommendations, but not the other three: eliminating debris shedding from the External Tank, hardening the Orbiter so it can better survive debris impacts, and developing an on-orbit method of repairing the shuttle's thermal protection system. Michael Griffin, who became NASA Administrator in April 2005, said that NASA and contractor personnel are those responsible and accountable for determining if and when the shuttle is ready for RTF, and would not commit to meeting every CAIB recommendation.

Return to Flight (RTF)

NASA launched the space shuttle *Discovery* on the first of two "Return to Flight" (RTF) missions — STS-114 — on July 26, 2005 and it successfully landed on August 9. On July 27, however, NASA announced that a piece of foam had detached from STS-114's ET during launch, similar to what happened to *Columbia.* Cameras and other sensors on *Discovery* and on the International Space Station — to which *Discovery* was docked for much of its mission — imaged the Orbiter and determined that it was not damaged, but further shuttle launches were suspended. Meanwhile, the images revealed that two "gapfillers" — ceramic coated fabric placed between thermal protection tiles —were protruding on the belly of the Orbiter that could have affected aerodynamic heating during reentry. One of the *Discovery* astronauts removed them during a space walk. The second RTF mission — STS-121 — was scheduled for September 2005, but deferred.

STS-121 launched on July 4, 2006, and returned safely to Earth on July 17. The shuttle *Atlantis* launched September 9 on STS 115, during which construction of the International Space Station was resumed. *Atlantis* returned to earth September 21. *Discovery* was launched on STS 116 December 9 and returned 13 days later on December 22. STS-117 was planned to launch in March 2007, but *Atlantis*, on the launch pad in a late February hailstorm, was so damaged that it had to be returned to the Vehicle Assembly Building for repairs. It is currently scheduled to launch in June or July.

Current plans for the shuttle include 16 flights (including STS 121, STS 115, and STS 116) to complete the ISS before the shuttle is permanently grounded in 2010. Also under consideration was another flight to service the Hubble Space Telescope. Following the Columbia disaster, then-Administrator Sean O'Keefe had cancelled the Hubble servicing mission, partly on the grounds that shuttle astronauts would not be able to reach the ISS as a haven in case the shuttle was unable to return to earth, but the decision was put under review by the new Administrator, Michael Griffin.

On October 31, 2006, NASA Administrator Griffin announced that the Hubble mission would be undertaken in May 2008. The servicing would extend the life of the telescope through 2013. To deal with emergencies, NASA plans to prepare a "launch on need" mission with a second shuttle ready to launch on a rescue mission if the first is found defective during the servicing mission.

The United Space Alliance (USA)

In 1995, NASA decided to turn most shuttle operations over to a "single prime contractor" — the United Space Alliance (USA), a limited liability company owned 50-50 by Boeing and Lockheed Martin. USA was created to pull together the 86 separate contracts with 56 different companies under which the shuttle program was then operating. NASA officials assert that it has saved $1 billion a year compared to what the costs would have been without it. NASA manages separate contracts with Lockheed Martin for the External Tank, ATK Thiokol for the Solid Rocket Boosters, and Pratt and Whitney Rocketdyne (owned by United Technologies) for the Space Shuttle Main Engines.

The Shuttle's Future

NASA attempted unsuccessfully for many years to develop a "second generation" reusable launch vehicle (RLV) to replace the shuttle. In 2002 NASA indicated the shuttle would continue flying until at least 2015, and perhaps 2020 or beyond. The *Columbia* tragedy, and President Bush's 2004 Vision for Space Exploration — to return astronauts to the Moon by 2020 and someday send them to Mars — forced NASA to revise that plan.

The President's Vision calls for the shuttle program, which absorbs approximately 25% of NASA's annual budget, to be terminated in 2010. A primary motivation is to make that funding available to implement other aspects of the Vision, although there also is concern about shuttle safety. Congress has been debating the Vision, including its impact on the shuttle and on U.S. human access to space. Some Members want to terminate the shuttle earlier than 2010 because they feel it is too risky and/or that the funds should be spent on accelerating the Vision. Others want to retain the shuttle at least until a new spacecraft, the Crew Exploration Vehicle (CEV), is available to take astronauts to and from the ISS. The CEV is now planned for 2012 at the earliest, leaving a multi-year gap during which U.S. astronauts would have to rely on Russia for access to the ISS.

At the beginning of 2005, NASA officials indicated 28 shuttle flights were needed to complete ISS construction. In the fall of 2005, NASA announced a new plan showing 18 shuttle missions to the ISS, and possibly one more to service the Hubble Space Telescope (see CRS Report RS21767, *Hubble Space Telescope: Should NASA Proceed with a Servicing Mission?* by Daniel Morgan). As part of its FY2007 budget request, NASA reduced to 16 the number of shuttle flights it now plans to the ISS (plus one possible flight to Hubble). The two that were cut were logistics flights taking cargo to the ISS. NASA is hoping that a commercial launch service provider will develop vehicles that can perform that task. The original versions of what became the 2005 NASA authorization act (P.L. 109-155) had conflicting language about the future of the shuttle. The original Senate bill (S. 1281) directed NASA not to terminate the shuttle until a replacement was available; the House version (H.R. 3070) directed NASA not to fly the shuttle after December 31, 2010. The final law states that it is U.S. policy to have human access to space on a continuous basis, and directs NASA to submit several related reports to Congress.

Shuttle Budget

Funding for the shuttle for FY2007 was not specified in P.L. 110-5, which appropriated a total of $6.140 billion for all space operations, including the shuttle and the ISS. For FY2008, NASA has requested $2.239 billion for the shuttle. (For details on the NASA budget, see CRS Report RS22625, *National Aeronautics and Space Administration: Overview, FY2008 Budget in Brief, and Key Issues for Congress*, by Daniel Morgan and Carl E. Behrens.)

ISSUES FOR CONGRESS

In passing the 2005 NASA authorization act (P.L. 109-105), Congress basically agreed with the President's plan for directing NASA's attention to a return to the Moon and manned missions to Mars. Included in the Moon-Mars "Vision" is the plan to end flights of the Space Shuttle in 2010, and restriction of U.S. experiments on the ISS mostly to those that forward the goal Moon-Mars goal. A number of critical questions remain, however.

- Adequacy of funding is the chief question raised about NASA's activities. In presenting the Moon-Mars vision, the President did not request significantly increased money for NASA, despite chronic indications that the missions it was already charged with were underfunded. NASA has responded to the new mission by cutting back funding for its other activities, primarily in scientific research and aeronautics.
- Although *Discovery's* "Return to Flight" mission of July 2006 was a success, the ability of the shuttle fleet to carry out enough flights to complete construction of the ISS by 2010 is still in question. With a history of more than a hundred successful missions, it might be assumed that another 15 or so would be considered more or less routine, but instead, each launch is still a major and risky event. The great complexity of the vehicle and the extreme environment in which it operates require constant attention to possible accidents and malfunctions, many of which must be addressed on an ad hoc basis.
- The future role of the ISS is also unclear. Assuming that enough shuttle flights are made to carry out "core completion" of the station by 2010, it is not clear what will be done with the ISS after that. In particular, there will be a gap of several years between retirement of the shuttle in 2010 and beginning of flight of the Crew Exploration Vehicle, to be designed for the return to the moon but able to serve as a vehicle to reach the ISS. The current schedule is to fly the CEV by 2012, but design of the vehicle is just beginning.

References

[1] National Aeronautics and Space Administration. Columbia Accident Investigation Board Report. August 2003. See CRS Report RS21606, NASA's Space Shuttle Columbia: Synopsis of the Report of the Columbia Accident Investigation Board, by Marcia S. Smith.

In: Space Policy and Exploration
Editor: W. N. Callmers, pp. 101-109
ISBN: 978-1-60456-448-8

Chapter 5

U.S. Military Space Programs: An Overview of Appropriations and Current Issues*

Patricia Moloney Figliola

Abstract

The 1958 National Aeronautics and Space Act specified that military space activities be conducted by the Department of Defense (DOD). DOD and the intelligence community manage a broad array of space activities, including launch vehicle development, communications satellites, navigation satellites (the Global Positioning System — GPS), early warning satellites to alert the United States to foreign missile launches, weather satellites, reconnaissance satellites, and developing capabilities to protect U.S. satellite systems and to deny the use of space to adversaries (called "space control" or "counterspace systems"). The 1990-1991 Persian Gulf War is dubbed by some as the first "space war" because support from space displayed great improvement over what was available during the previous major conflict, Vietnam. These systems continue to play significant roles in U.S. military operations. How to organize DOD and the intelligence community to work effectively on space programs has been an issue for many years.

Tracking the DOD space budget is extremely difficult since space is not identified as a separate line item in the DOD budget. Additionally, DOD sometimes releases only partial information (omitting funding for classified

* Excerpted from CRS Report RL33601, dated August 7, 2006.

programs) or will suddenly release without explanation new figures for prior years that are quite different from what was previously reported. Figures provided to CRS show a total (classified and unclassified) DOD space budget of $19.4 billion for FY2003, $20 billion for FY2004, $19.8 billion for FY2005, and a request of $22.5 billion for FY2006. The actual FY2006 and proposed FY2007 budget figures are not yet available.

Two DOD space programs that have been particularly controversial are Space Radar (formerly Space-Based Radar — SBR) and TSAT (the transformational communications satellite program). The programs are controversial because their cost estimates are high, and Congress has been skeptical of those estimates and of DOD's ability to manage the programs successfully based on past program performance. Congress cut DOD's $226 million FY2006 request for Space Radar by $126 million and its $836 million FY2006 request for TSAT by $400 million. The FY2007 requests for those programs are $266 million for Space Radar and $867 million for TSAT.

BACKGROUND

The 1958 National Aeronautics and Space Act specified that military space activities be conducted by the Department of Defense (DOD). The Undersecretary of the Air Force is DOD's executive agent for space. The intelligence community makes significant use of space-based intelligence collection capabilities. The National Reconnaissance Office (NRO), an agency within DOD, builds and operates intelligence-collection satellites and collects and processes the resulting data, which are provided to users such as the National Geospatial-Intelligence Agency (NGA) and the National Security Agency (NSA). NRO, NGA, and NSA are all under the oversight of the new Director of National Intelligence (DNI).[1]

DOD and the intelligence community manage a broad array of space activities, including launch vehicle development, communications satellites, navigation satellites (the Global Positioning System — GPS),[2] early warning satellites to alert the United States to foreign missile launches, weather satellites, reconnaissance satellites, and developing capabilities to protect U.S. satellite systems and to deny the use of space to adversaries (called "space control" or "counterspace systems"). The 1990-1991 Persian Gulf War is dubbed by some as the first "space war" because support from space displayed great improvement over what was available during the previous major conflict, Vietnam. These systems continue to play significant roles in U.S. military operations.

How to organize DOD and the intelligence community to work effectively on space programs has been an issue for many years. Congress established commissions to review the NRO in the FY2000 intelligence authorization act, P.L. 106-120; NGA (then called NIMA, the National Imagery and Mapping Agency) in the classified annex to the FY2000 DOD appropriations act, P.L. 106-79; and overall U.S. national security space management and organization in the FY2000 DOD authorization act, P.L. 106-65. The NRO, NGA/NIMA, and "Rumsfeld Space Commission" reports are discussed below.

Although U.S. military and civilian space programs are separated organizationally, the functions performed by satellites and the vehicles that launch them are not easily divided. Both sectors use communications, navigation, weather, and remote sensing/reconnaissance satellites, which may operate at different frequencies or have different capabilities, but have similar technology. The same launch vehicles can be used to launch any type of military, civilian, or commercial satellite. DOD uses some civilian satellites and vice versa.

After the Cold War, interest in space weapons to attack satellites (antisatellite, or ASAT, weapons) or ballistic missiles declined initially, but was rekindled beginning with the 104th Congress. Using satellites to attack ballistic missiles has been controversial since President Reagan's 1983 announcement of a Strategic Defense Initiative to study the viability of building a ballistic missile defense system to protect the United States and its allies. The Clinton Administration changed the name of the Strategic Defense Initiative Organization to the Ballistic Missile Defense Organization to reflect a new focus on theater missile defense in the wake of the Persian Gulf War, rather than national missile defense. The George W. Bush Administration changed the name to the Missile Defense Agency (MDA) to reflect its interest in broad missile defense goals.[3]

The concept of placing weapons in space, as part of a missile defense system or otherwise, remains controversial. A May 18, 2005, *New York Times* article reported that the new national space policy being developed by the Bush Administration would "move the United States closer to fielding offensive and defensive space weapons."[4] Then-White House Press Secretary Scott McClellan, responding to questions at a White House press briefing, stressed that the new policy, still being developed, does not represent a substantial shift in U.S. policy. The same day, Representative Kucinich introduced a bill (H.R. 2420) to ban weapons in space and the use of such weapons to damage or destroy objects in orbit. The House rejected (124-302) a Kucinich amendment to the Foreign Relations Authorization Act (H.R. 2601) on July 20, 2005, that was similar to his bill.

DOD Space Budget

Space is not a line item in the DOD budget and DOD's annual budget justifications do not include a figure for "space activities"; therefore, DOD funding figures must be used cautiously. DOD sometimes releases only partial information or will release without explanation new figures for prior years that are quite different from what was previously reported.[5]

According to the data provided by the DOD Office of the Comptroller in Spring 2005, DOD requested $22.5 billion for space programs for FY2006. Newer figures showing what Congress appropriated for FY2006 and the FY2007 request are not yet available.

FY2006 Authorization and Appropriations

The FY2006 authorization and appropriations bills, listed below, contain the authority and funding for DOD space activities, but, as mentioned, do not specify figures for those activities.

P.L. 109-148, H.R. 2863. FY2006 DOD appropriations bill. H.R. 2863 reported from House Appropriations Committee June 10, 2005 (H.Rept. 109-119); passed House June 20. Reported from Senate Appropriations Committee September 29 (S.Rept. 109-141); passed Senate October 7. Conference report (H.Rept. 109-359) passed House December 19. Signed into law (P.L. 109-148) December 30, 2005.

P.L. 109-163, H.R. 1815. FY2006 DOD authorization bill. H.R. 1815 reported from House Armed Services Committee May 20, 2005 (H.Rept. 109-89); passed House May 25. S. 1042 reported from Senate Armed Services Committee May 17 (S.Rept. 109-69); passed Senate November 15. Conference report (H.Rept. 109-360) passed House December 19, Senate December 21. Signed into law (P.L. 109-163) January 6, 2006.

MILITARY SPACE PROGRAM ISSUES

For many years, questions have arisen about whether DOD effectively manages its space activities, and several commissions and task forces have studied the issue. Congress created a commission in the FY2000 DOD authorization bill to make recommendations on the overall management of national security space programs. Chaired by Donald Rumsfeld, the commission released its report on January 11, 2001, shortly after Mr. Rumsfeld became Secretary of Defense. The "Rumsfeld Space Commission" made sweeping recommendations for management of DOD and intelligence community space programs. According to two GAO reports, DOD intended to implement 10 of the 13 organizational recommendations,[6] although no additional updates have been provided.

Several DOD space programs have experienced significant cost overruns and schedule delays, raising concerns about DOD's acquisition process for space systems.

The Defense Science Board (DSB) and Air Force Scientific Advisory Board (AFSAB) commissioned a task force chaired by retired Lockheed Martin executive Tom Young to review DOD space program acquisition because of significant cost increases in several programs; its May 2003 report was publicly released in September 2003.[7] Four key findings of the report were that cost has replaced mission success as the primary driver in managing acquisition processes, creating excessive technical and schedule risk; the space acquisition system is strongly biased to produce unrealistically low cost estimates; government capabilities to lead and manage the acquisition process have seriously eroded; and there are long term concerns about the space industrial base. According to press reports, the task force produced an update in August 2004 that concluded that some of the space programs it criticized were making progress but still required close review, and that better coordination is needed between the military and intelligence agencies in setting requirements.[8]

On April 6, 2006, the Senate Committee on Armed Forces held a hearing on space acquisitions. At that hearing, Cristina T. Chaplain, GAO's Acting Director of Acquisition and Sourcing, testified that DOD's space acquisition programs continue to face substantial cost and schedule overruns. In some cases, according to Ms. Chaplain, cost growth has come close to or exceeded 100%, causing DOD to nearly double its investment with no corresponding increase in functionality. Additionally, many programs have experienced significant schedule delays — as much as six years — postponing delivery of promised capabilities to the warfighter.[9]

SBIRS-High

DOD is developing a new satellite system to replace its Defense Support Program series of early warning satellites that alert the National Command Authority to foreign missile launches. Called SBIRS-High (Space-Based Infrared System-High), it has encountered significant schedule delays and cost growth, breaching "Nunn-McCurdy" cost-growth limits several times.[10] The May 2003 report of the Defense Science Board and Air Force Scientific Advisory Board criticized early program management of SBIRS-High, and took a cautious attitude concerning whether the restructured program would succeed.[11] An October 2003 GAO report[12] concluded the program remained at "substantial risk of cost and schedule increases."

SBIRS-High was designed as a constellation of five satellites above the equator in geostationary orbit (GEO) plus sensors on two other satellites in highly-elliptical orbits (HEO). DOD still plans to launch the sensors on the two HEO satellites, but will procure, at most, three of the GEO satellites. The funds that would have been spent for the fourth and fifth GEO satellites reportedly will be used instead to design an alternative system using state-of-the-art technologies. DOD's Selected Acquisition Report for the quarter ending September 2005[13] showed the original cost estimate for SBIRS-High as $3.68 billion in FY1995 dollars (over $4 billion in current year dollars, i.e., adjusted for inflation), compared with a new estimate of $9.01 billion in FY1995 dollars (over $10 billion in current year dollars). The FY2007 request for SBIRS-High is $669 million.

Space Radar and TSAT

DOD's requests to initiate new programs, including a Space Radar program (previously called Space-Based Radar), and the Transformational Satellite (TSAT) communications satellite program, are controversial because of the potentially large costs involved, questions concerning whether the technologies they require are sufficiently mature, and issues regarding ways to avoid the cost growth and schedule delays experienced in other DOD space programs.

Space Radar

Space Radar would be a system of many satellites (the exact number has not been determined) that would track mobile targets (as opposed to fixed targets) on the ground. The House Appropriations Committee has sharply criticized the program for the past several years. In its 2004 report on the FY2005 DOD appropriations bill (H.Rept. 108-557), the committee noted that the estimated cost for a nine-satellite constellation was $34 billion, and the Air Force considers nine satellites to be less than half the number required. The committee expressed skepticism about the $34 billion estimate, as well.

The FY2006 request for Space Radar was $226 million. Congress cut $126 million in both the FY2006 DOD authorization and appropriations acts. The Senate Armed Services Committee (SASC) and House Armed Services Committee (HASC) commented extensively on the program (pp. 200-201 of S.Rept. 109-69; pp. 214-216 of H.Rept. 109-89). Both discussed the need to integrate the Space Radar into a broader architecture of radar capabilities, including airborne radars. SASC also emphasized the need for a single space radar system to meet military and intelligence needs, and expressed concern about the lack of certainty about cost and other issues. HASC noted a number of concerns, including that the Air Force has not sufficiently emphasized affordability as a key objective or fully considered the requirements for a demonstration program. HASC provided specific direction as to what is needed for such a program, utilizing ground, airborne, and existing space assets.

TSAT

The TSAT program would be a follow-on to the Advanced Extremely High Frequency (AEHF) program, which, in turn, is a follow-on to the current series of Milstar satellites. AEHF itself is controversial because of cost overruns, and, in 2002, DOD decided to procure only three instead of five AEHF satellites. The first AEHF launch is scheduled for 2008. TSAT is expected to "transform" DOD communications by providing vastly greater capacity than is available today by operating at much higher (optical) frequencies. If TSAT is delayed, some observers suggest that additional AEHF satellites may be needed.

In May 2006, GAO released a report outlining the ongoing issues and problems, in the development and deployment of the TSAT system.[14] Specifically, GAO stated that DOD was not meeting original cost, schedule, and performance goals established for the TSAT program. However, GAO noted that

DOD is taking positive steps to lower risk in the TSAT program so it can enter the product development phase with greater chance of success.

The FY2006 request for TSAT was $836 million. Congress cut $400 million in the FY2006 DOD authorization and appropriations acts. In its report on the bill (S.Rept. 109-69), SASC expressed support for TSAT, but noted that GAO found that only one of its seven critical technologies is mature (p. 200). Thus, SASC recommended that a fourth AEHF satellite be procured (adding $100 million for that satellite) while the TSAT technologies are developed, and that some of those technologies could be incorporated into the fourth AEHF satellite if feasible. The Senate Appropriations Committee did not add funding for a fourth EHF satellite, but restricted the use of $150 million of the funds it provided for TSAT in order to fund a fourth EHF satellite if needed (S.Rept. 109-141, p. 218). HASC also supported the concept of TSAT, and commended the Air Force on its vision for revolutionary solutions (pp. 216-217). However, it cited the recent history of cost overruns and schedule growth associated with other Air Force space programs as cause for skepticism about the ability of the current acquisition system to accommodate the risks associated with revolutionary technologies. The HASC directed that the focus of the TSAT program shift to technology development rather than acquisition. It added that it did not believe additional funding for AEHF would be needed until FY2007.

REFERENCES

[1] See CRS Report RL32515, *Intelligence Community Reorganization: Potential Effects on DOD Intelligence Agencies*, by Richard A. Best, Jr., for more on the DNI and potential effects for DOD intelligence agencies, including NRO, NGA, and NSA.

[2] For additional information on GPS, see "The Future of the Global Positioning System," Defense Science Board, October 2005, online at [http://www.acq.osd.mil/dsb/reports /2005-10-GPS_Report_Final.pdf].

[3] See CRS Report RL31111, *Missile Defense: The Current Debate*, coordinated by Steven A. Hildreth.

[4] Weiner, Tim, "Air Force Seeks Bush's Approval for Space Arms," *The New York Times*, May 18, 2005, p1.

[5] Space spending by all federal government agencies, by year since FY1959, is provided in Appendix E of the annual Aeronautics and Space Report of the President, submitted to Congress by NASA. The most recent edition [http://history.nasa.gov/presrep2004.pdf] covers funding through FY2004.

[6] "Defense Space Activities: Status of Reoganization," GAO-02-772, June 2002, and "Defense Space Activities: Organizational Changes Initiated, but Further Management Actions Needed," GAO-03-379, April 2003.

[7] Report of the Defense Science Board/Air Force Scientific Advisory Board Joint Task Force on Acquisition of National Security Space Programs, Office of the Under Secretary of Defense for Acquisition, Technology, and Logistics, May 2003, online at [http://www.acq.osd.mil/dsb/ reports/ space.pdf].

[8] See, for example, Merle, Renae, "Report Says Air Force's Space Programs Improved," *The Washington Post*, August 25, 2004, E02.

[9] Ms. Chaplain's full testimony is available online at [http://armed-services.senate.gov/e_witnesslist.cfm?id=1823]. Additional written comments submitted in response to specific questions from the committee are available online at [http://www.gao.gov/new.items/d06776r.pdf].

[10] 10 U.S.C § 2433, commonly referred to as the "Nunn-McCurdy" provisions, requires the Pentagon to justify continuation of a program whose costs have grown by 25% or cancel the project.

[11] Report of the Defense Science Board/Air Force Scientific Advisory Board Joint Task Force on Acquisition of National Security Space Programs, Office of the Under Secretary of Defense for Acquisition, Technology, and Logistics, May 2003, online at [http://www.acq.osd.mil/dsb/reports/ space.pdf].

[12] "Despite Restructuring, SBIRS High Program Remains at Risk of Cost and Schedule Overruns," GAO-04-48, October 2003.

[13] Selected Acquisition Report for Quarter ending September 2005, Office of the Secretary of Defense, online at [http://www.acq.osd.mil/ara/am/sar/ 2005-SEP-SST.pdf].

[14] Space Acquisitions: DOD Needs Additional Knowledge as it Embarks on a New Approach for Transformational Satellite Communications System, GAO, May 2006, available online at [http://www.gao.gov/new.items/ d06537.pdf].

In: Space Policy and Exploration
Editor: W. N. Callmers, pp. 111-119

ISBN: 978-1-60456-448-8

Chapter 6

NASA'S SPACE SHUTTLE PROGRAM: THE *COLUMBIA* TRAGEDY, THE *DISCOVERY* MISSION, AND THE FUTURE OF THE SHUTTLE*

Marcia S. Smith

ABSTRACT

On August 9, 2005, the space shuttle *Discovery* successfully completed the first of two "Return to Flight" (RTF) missions — STS-114. It was the first shuttle launch since the February 1, 2003, *Columbia* tragedy. NASA announced on July 27, 2005, the day after STS-114's launch, that a second RTF mission would be indefinitely postponed because of a problem that occurred during *Discovery*'s launch that is similar to what led to the loss of *Columbia*. The next launch is currently expected some time in 2006. This chapter discusses the *Columbia* tragedy, the *Discovery* mission, and issues for Congress regarding the future of the shuttle. For more information, see CRS Issue Brief IB93062, *Space Launce Vehicles: Government Activities, Commercial Competition, and Satellite Exports*, by Marcia S. Smith.

* Excerpted from CRS Report RS21408, dated January 4, 2006.

THE LOSS OF THE SPACE SHUTTLE *COLUMBIA*

The space shuttle *Columbia* was launched on its STS-107 mission on January 16, 2003. After completing a 16-day scientific research mission, *Columbia* started its descent to Earth on the morning of February 1, 2003. As it descended from orbit, approximately 16 minutes before its scheduled landing at Kennedy Space Center, FL, *Columbia* broke apart over northeastern Texas. All seven astronauts aboard were killed: Commander Rick Husband; Pilot William McCool; Mission Specialists Michael P. Anderson, David M. Brown, Kalpana Chawla, and Laurel Clark; and payload specialist Ilan Ramon, an Israeli. The last communication with *Columbia* was at about 09:00 EST. The shuttle was at an altitude of 207,135 feet, traveling at a speed of Mach 18.3 (about 13,000 miles per hour).

THE SPACE SHUTTLE *COLUMBIA* AND THE STS-107 MISSION

The Space Transportation System (STS) — the space shuttle — consists of an airplane-like orbiter, two Solid Rocket Boosters (SRBs) on either side, and a large cylindrical External Tank that holds the fuel for the orbiter's main engines. The SRBs detach from the orbiter 2½ minutes after launch when their fuel is spent, fall into the ocean, and are recovered for refurbishment and reuse. The External Tank is not reused, but is jettisoned as the orbiter reaches Earth orbit, and disintegrates as it falls into the Indian Ocean. At the beginning of 2003, all planned shuttle launches, other than STS-107 and two missions to the Hubble Space Telescope, were scheduled to support assembly and operation of the International Space Station (see CRS Issue Brief IB93017).

Columbia was one of four spaceflight-worthy reusable space shuttle orbiters in NASA's fleet. The remaining orbiters are *Discovery*, *Atlantis*, and *Endeavour*. A fifth orbiter, *Challenger*, was lost in a 1986 accident. Another orbiter, *Enterprise*, was used for approach and landing tests in the 1970s and was not designed to travel in space. *Enterprise* now belongs to the Smithsonian's National Air and Space Museum.

Columbia was the first spaceflight-worthy orbiter built for NASA by Rockwell International (the space division of Rockwell, which built the orbiters, was later bought by Boeing). It was used for the very first shuttle flight on April 12, 1981. The STS-107 mission was *Columbia*'s 28th flight. Although *Columbia* was the oldest orbiter, *Discovery* has been used for more flights (30). NASA has conducted a total of 114 shuttle launches to date. Orbiters are periodically taken out of service for maintenance and overhaul. *Columbia* had lost undergone such an "orbiter major modification" (OMM) period in 1999-2001. STS-107 was *Columbia*'s second flight after the OMM. It was a scientific research mission that, unlike most current shuttle launches, was not related to the International Space Station (ISS) program (see CRS Issue Brief IB93017). The crew conducted a research program involving 59 separate investigations. Some of the research required analysis of specimens and data sets after the shuttle returned to Earth, and most were destroyed along with the crew and orbiter. Other data, however, were transmitted to ground-based researchers during the flight, and a few specimens were retrieved among the debris, so some of the research survived. Quantifying the amount is difficult.

PREVIOUS SPACEFLIGHT-RELATED CREW FATALITIES

The United States has suffered two other spaceflight-related accidents that caused astronaut fatalities. On January 27, 1967, the crew of the first Apollo mission — Virgil "Gus" Grissom, Edward White, and Roger Chaffee — died when electrical arcing in spacecraft wiring caused a fire in their Apollo command module during a pre-launch test. Apollo flights resumed after 21 months. On January 28, 1986, the space shuttle *Challenger* (STS 51-L) exploded 73 seconds after launch, killing all seven astronauts aboard: Francis "Dick" Scobee, Michael Smith, Judith Resnik, Ellison Onizuka, Ronald McNair, Gregory Jarvis (a payload specialist from Hughes Aircraft), and schoolteacher Christa McAuliffe. A commission chaired by former Secretary of State William Rogers determined that cold weather at the launch site caused a rubber "O-ring" in one of the SRBs to fail, allowing gases to escape, resulting in a catastrophic explosion. The shuttle system was grounded for 32 months.

Four Soviet cosmonauts died during spaceflights. Cosmonaut Vladimir Komarov died during the first Soyuz flight on April 24, 1967. The spacecraft's parachutes did not function properly and it struck the ground with great force, killing Colonel Komarov. Soviet human spaceflights were suspended for 18 months. Three cosmonauts died on Soyuz 11 on June 29, 1971 when an improperly sealed valve allowed the spacecraft's atmosphere to vent into space. The cosmonauts — Georgiy Dobrovolskiy, Vladislav Volkov, and Viktor Patsayev — were not wearing spacesuits, and were asphyxiated. There were no Soviet human spaceflights for 27 months

THE *COLUMBIA* ACCIDENT INVESTIGATION BOARD (CAIB)

Then-NASA Administrator Sean O'Keefe established the *Columbia* Accident Investigation Board (CAIB) within hours of the tragedy, and transitioned responsibility for the investigation to it on February 6, 2003. Chaired by Adm. (Ret.) Harold Gehman, former NATO Supreme Allied Commander, Atlantic, CAIB had 12 other members (see [http://www.caib.us]). All were appointed by Mr. O'Keefe, although some were added to the initial roster upon the recommendation of Adm. Gehman. NASA revised the Board's charter three times to clarify its independence from NASA, primarily in response to congressional concerns. However, the CAIB was created by NASA, included NASA representatives, and the Board members were appointed by the NASA Administrator, so concerns about its independence continued. Title VIII of the FY2007-2008 NASA authorization act (P.L. 109-155) specifies how future human space flight accidents are to be investigated. CAIB released the results of its investigation on August 26, 2003 in Vol. 1 of its report; Volumes II-VI were released in October 2003. All are available at CAIB's website. Board member Brig. Gen. Duane Deal wrote a 10-page supplement, published in Vol. 2, providing additional recommendations and viewpoints.

The Causes of the Accident

The Board concluded that the tragedy was caused by both technical and organizational failures. The technical cause was damage to *Columbia*'s left wing by a 1.7 pound piece of insulating foam that separated from the External Tank's left "bipod ramp" and struck the orbiter's left wing 81.9 seconds after launch. The foam strike created a hole in a Reinforced Carbon-Carbon (RCC) panel on the leading edge of the wing, allowing superheated air (perhaps exceeding 5,000^{o}F) to enter the wing during reentry. The extreme heat caused the wing to fail structurally, creating aerodynamic forces that led to the disintegration of the orbiter. Organizationally, the Board pointed to detrimental cultural traits and organizational practices that developed over the institutional history of the program. Adm. Gehman cited a loss of "checks and balances" in the program's management that should have led to a recognition of the danger posed by "foam shedding" from the External Tank, which had occurred on previous shuttle missions. The Board also cited long term budget constraints, and schedule pressure associated with assembly of the International Space Station, as factors.

CAIB's Recommendations and the Stafford/Covey Task Group

The CAIB made 29 recommendations, of which 23 were technical and six were organizational. Of the 29 recommendations, the Board specified that 15 be completed before the shuttle returned to flight status (see CRS Report RS21606 for a synopsis).

NASA created a task group chaired by two former astronauts — Thomas Stafford and Richard Covey — to evaluate NASA's compliance with the 15 CAIB recommendations that were to be completed before Return to Flight (RTF). The Stafford/Covey Task Group ([http://returntoflight.org]) did not address organizational or cultural issues, and was not tasked to determine whether the shuttle was ready to fly. The Task Group held its final public meeting on June 27, 2005, concluding that NASA had met the intent of 12 of the 15 CAIB RTF recommendations, but not the other three —preventing debris shedding from the External Tank, hardening the orbiter against damage from debris, and developing inspection and repair techniques for the thermal protection system. Some Task Group members indicated, however, that NASA had made considerable improvements to shuttle safety, and considered the shuttle ready to fly.

Then-NASA Administrator O'Keefe said during 2003 and 2004 that not only would NASA comply with the CAIB recommendations, but would "raise the bar" to ensure the shuttle is as safe as possible. Dr. Michael Griffin became NASA Administrator in April 2005. At an April 18 press conference, he said he would listen carefully to advice from the Stafford/Covey group, but that NASA and contractor personnel were those responsible and accountable for determining if and when the shuttle is ready for RTF. He would not commit to meeting every CAIB recommendation. In testimony to the House Science Committee on June 28, 2005, he said that NASA had spent lot of money trying to meet all the recommendations, but did not have the knowledge to do so in every case, such as stopping all debris shedding from the External Tank.

The CAIB report also cited cultural issues (e.g., the safety culture) that could affect shuttle safety. Those issues were addressed by an internal NASA study led by then-NASA Goddard Space Flight Center Director Al Diaz. NASA also hired an outside consulting firm, BST, to assess NASA culture and make recommendations about what changes were needed. A February 5, 2005 BST report concluded that there was improvement, but a significant number of people at NASA's field centers did not yet perceive change. Dr. Griffin terminated the contract with BST soon after becoming Administrator, saying that NASA could deal with these issues internally.

RETURN TO FLIGHT: THE STS-114 (*DISCOVERY*) MISSION

NASA launched the space shuttle on its first Return to Flight (RTF) mission, STS-114, on July 26, 2005, using the *Discovery* orbiter. The STS-114 crew was composed of six NASA astronauts and one from the Japan Aerospace Exploration Agency (JAXA): Commander Eileen Collins, Pilot James Kelly, and Mission Specialists Charles Camarda, Wendy Lawrence, Soichi Noguchi (JAXA), Steve Robinson, and Andy Thomas. STS-114 docked with the International Space Station (ISS) for most of its mission. STS-114 delivered supplies, repaired ISS components, and tested inspection and repair techniques for the orbiter. The STS-114 mission was extended one day so the shuttle crew could assist the ISS crew in stowing supplies. Inclement weather at Kennedy Space Center, FL delayed landing until August 9, and diverted it to Edwards Air Force Base, CA.

The day after launch (July 27), NASA announced that further shuttle launches would be indefinitely postponed after discovering that a comparatively large piece of foam broke off from *Discovery*'s External Tank during launch, similar to what happened to *Columbia.* In this case, NASA determined that it did not hit the orbiter. Subsequently, several smaller pieces were found to have broken off and may have hit the orbiter's wing. Cameras and other sensors aboard *Discovery* and the ISS took detailed images to determine if there was any damage and none was discovered. However, the images revealed that two "gapfillers" — ceramic-coated fabric that fills gaps between the thermal protection tiles — on the bottom of the orbiter were protruding, which could affect aerodynamic heating during reentry. Astronaut Robinson removed them during a spacewalk.

The second RTF mission was scheduled for September 2005 with the mission of continuing testing post-*Columbia* fixes and resupplying the ISS. NASA continues to study why the foam loss on STS-114 occurred. That work was slowed by hurricane damage to the Michoud Assembly Facility, in New Orleans, LA, where the External Tanks are manufactured. It is operated for NASA by Lockheed Martin. NASA's Stennis Space Center, near Slidell, LA, which tests the space shuttle main engines, also suffered hurricane damage. The next shuttle launch is expected no earlier than May 2006.

ISSUES FOR CONGRESS

Cost

In July 2004, NASA released a cost estimate for RTF covering the years FY2003-2009: $2.2 billion, double the previous estimate of $1.1 billion. For FY2005, NASA requested $4.3 billion for the shuttle program. In November 2004, NASA informed Congress it needed $762 million more than expected for FY2005. In the FY2005 Consolidated Appropriations Act (P.L. 108-447), Congress approved the $4.3 billion requested, subject to an across-the-board 0.80% rescission, and said that NASA could submit a request for supplemental appropriations, or reprogram funds from other NASA programs. According to a May 10, 2005 update to its FY2005 operating plan, NASA reprogrammed the following funds into RTF: $55 million from the Science Mission Directorate ($20 million from space science, $35 million from earth science); $375.8 million from the Exploration Systems Mission Directorate ($73 million from biological and physical research, $204 million from human and robotic technology, and $98 million from transportation systems); and $331.2 million from the Space Operations Mission Directorate ($160 million from the space station, $170 million from space shuttle upgrades, and $1.2 million from space flight support). Separately, Congress appropriated $126 million to NASA in an FY2005 emergency supplemental (P.L. 108-324) for damages at Kennedy Space Center, FL, from the 2004 Florida hurricanes. NASA officials say those funds are not available for RTF.

The CAIB noted that long term budget constraints were a factor in the *Columbia* tragedy. NASA requested and received $4.5 billion for the shuttle in the FY2006 appropriations bill that includes NASA (P.L. 109-108), subject to a 0.28% rescission in that act, and a 1% rescission in another appropriations act (P.L. 109-148). The latter act also included $350 million for NASA for recovery from Hurricane Katrina. The FY2006 request for the shuttle, developed by then-Administrator O'Keefe, included an "out-year projection" that the shuttle budget would decline to $2.4 billion by FY2010. Many observers considered that unrealistic, and at a November 3, 2005 House Science Committee hearing, Dr. Griffin agreed that NASA will need $3-$5 billion more than that for FY2008-FY2010. The source of those funds is unclear.

The Shuttle's Future

The foam-shedding event during *Discovery*'s launch and the resulting decision to suspend further shuttle launches, plus the funding shortfall in NASA's out-year projection, adds to the complexity of decisions about the shuttle's future. Prior to *Discovery*'s launch, discussion was focused on President Bush's January 2004 directive that the shuttle be retired as part of his new "Vision for Space Exploration" where NASA is to focus its attention on returning astronauts to the Moon by 2020 and someday sending them to Mars (see CRS Report RS21720). Under the plan, the shuttle — in its current form — would be terminated in 2010, primarily so that its funding can be redirected towards achieving other aspects of the Vision. The 2010 date also coincides with the CAIB's recommendation that if the shuttle is to be flown longer, it should be recertified. Two key issues raised by the decision to end shuttle flights by 2010 are the extent to which the United States wants its own ability to launch astronauts into space, and the importance of completing ISS construction and meeting U.S. commitments to other ISS partners.

The shuttle is the only U.S. vehicle capable of taking astronauts to space today. As part of the Vision, NASA is building a new "Crew Exploration Vehicle" (CEV). The original schedule called for it to be available for launching crews to Earth orbit by 2014, meaning there would be a multi-year gap between the end of the shuttle program and the availability of CEV. During that time, NASA would be dependent on Russia for human access to space. Dr. Griffin wants to accelerate CEV availability to reduce the gap as much as possible in part because he believes the United States should not be dependent on any other nation for human access to space. NASA's current target is 2012. The Senate adopted an amendment to the FY2006 Department of Defense authorization bill (S. 1042) on November 15, 2005 expressing the sense of the Senate that it is in the national security interest of the United States to maintain preeminence in human spaceflight. Initial versions of the NASA authorization bill (H.R. 3070/S. 1281) contained conflicting language about the future of the shuttle. The Senate bill originally directed NASA not to terminate the shuttle until a replacement was available, while the House version originally directed NASA not to fly the shuttle after December 31, 2010. The final version of the act (P.L. 109-155) states that it is U.S. policy to have human access to space on a continuous basis, and directs NASA to submit several related reports to Congress.

The shuttle's primary purpose today is assembly and servicing of the ISS. Dr. Griffin also promised to reassess whether or not to use the shuttle for one more servicing mission to the Hubble Space Telescope (see CRS Report RS21767). NASA's current target is to launch the shuttle 18 times to support the ISS, with one potential additional launch to Hubble, by the end of 2010. The need to complete ISS assembly and meet U.S. commitments to its ISS partners, and to service Hubble, is being weighed against the costs operating the shuttle for that many more years, especially now that NASA has conceded that it will need $3-5 billion more than was included in its out-year funding profile. One consideration is that some of the ISS segments waiting to be launched are those built by the other ISS partners, and President Bush said the United States would meet its commitments to those partners. Another question is whether setting a firm deadline for ending shuttle flights places undue schedule pressure on shuttle program managers to complete as many launches as possible within whatever remaining time there is. CAIB cited schedule pressure as a factor in the *Columbia* accident.

The effect on ISS utilization by NASA and its partners is another issue. ISS was designed to be serviced by the shuttle, which can carry larger crews than the Russian Soyuz spacecraft, and larger and heavier cargo than the Russian Progress cargo spacecraft. In addition, the shuttle is the only cargo spacecraft that can return large amounts of material to Earth (Russia's Progress burns up as it reenters the atmosphere). Without the shuttle, the results of scientific experiments, equipment needing repair, and other items could only be returned if they fit within the small confines of the Soyuz capsule along with whatever crew members were returning to Earth (unless new cargo spacecraft are developed). How NASA can ensure that the shuttle system will remain safe as workers and vendors move on to other projects as the shuttle program comes to an end is also being discussed. NASA plans to develop two "shuttle-derived" launch vehicles (see CRS Issue Brief IB93062), which could mitigate some of the shuttle workforce displacements.

In: Space Policy and Exploration
Editor: W. N. Callmers, pp. 121-130

ISBN: 978-1-60456-448-8

Chapter 7

SPACE EXPLORATION: ISSUES CONCERNING THE "VISION FOR SPACE EXPLORATION"*

Marcia S. Smith

ABSTRACT

On January 14, 2004, President George W. Bush announced new goals for the National Aeronautics and Space Administration (NASA), directing the agency to focus on returning humans to the Moon by 2020, and eventually sending them to Mars and "worlds beyond." The President invited other countries to join. Most of the funding for this "Vision for Space Exploration" is to be redirected from other NASA activities, including terminating the space shuttle program in 2010, and ending U.S. participation in the International Space Station by 2016. NASA released an implementation plan for the Vision on September 19, 2005, and estimated the cost of returning humans to the Moon by 2018 (NASA's current goal) at $104 billion. An estimate for sending people to Mars was not provided. This chapter identifies issues Congress has been considering as it debates the President's Vision. See CRS Issue Brief IB92011, *U.S. Space Programs: Civilian, Military, and Commercial*, by Marcia S. Smith, for further information.

---

* Excerpted from CRS Report RS21720, dated January 4, 2006.

Overview of President George W. Bush's Vision for Space Exploration

On January 14, 2004, President George W. Bush announced a new "Vision for Space Exploration" [http://www.whitehouse.gov/news/releases/ 2004/01/ 20040114-3.html]. Amplified by documents from the White House Office of Science and Technology Policy [http://www.ostp.gov] and NASA, the main features include the following.

- Astronauts would return to the Moon in the 2015-2020 time period using a new "Crew Exploration Vehicle" (CEV) to be developed by NASA. (The last time Americans walked on the Moon was in 1972.)
- Eventually, astronauts would go to Mars. No date was announced.
- Robotic probes would serve as trailblazers for human explorers. The first probe to support the Vision would be launched to the Moon in 2008.
- Construction of the International Space Station (ISS) would be completed by 2010. The President promised that the United States would meet its obligations to its partners in the ISS program — Europe, Canada, Japan, and Russia.[1] U.S. involvement in ISS would end by 2016.
- The space shuttle system would be retired in 2010. Between 2010 and 2014, when the CEV would be operational for Earth orbital missions, U.S. astronauts would rely on Russia to take them to and from ISS.
- NASA would redirect its research aboard the ISS to that which specifically supports the Vision, instead of the broadly based research program that was planned.
- Other countries were invited to participate in the Vision.
- A cost estimate was not provided, but FY2005 NASA budget documents showed that $12.6 billion would be "added"[2] to its budget for FY2005-2009 to begin implementing the Vision, and a NASA projected budget chart (the "sand chart"[3]) suggested that $150-170 billion would be spent on the Vision from FY2004-2020. Most of the money was to come from other NASA programs. The $12.6 billion, for example, comprised $1 billion in new money, and $11.6 billion redirected from other activities.
- A Commission on the Implementation of U.S. Space Exploration Policy was created by the President to advise NASA on implementation of the Vision; its report was issued in June 2004.

The speech came 11 months after the February 1, 2003, space shuttle *Columbia* accident (see CRS Report RS21408), and two weeks after the January 3, 2004, successful landing of a U.S. robotic probe (Spirit) on Mars. Invoking the explorations of Lewis and Clark, the President explained that America has ventured into space for the same reasons, "because the desire to explore and understand is part of our character. And that quest has brought tangible benefits that improve our lives in countless ways."

Public Reaction

An Associated Press-Ipsos poll conducted before the President's speech, in response to press reports that it was imminent, found 48% of the respondents in favor of a Moon/Mars program, 48% opposed, and 4% not sure.[4] A June 22-July 7, 2004 Gallup poll sponsored by the Coalition for Space Exploration (a group of companies and organizations that support the Vision) found that, in response to a question about funding a Moon/Mars program as long as NASA's budget did not exceed 1% of the federal budget, 26% strongly supported, 42% supported, 15% opposed, and 9% strongly opposed the plan. A similar poll a year later found that 26% strongly supported, 50.5% supported, 14% opposed, and 6.5% strongly opposed it. The remainder were neutral, did not know, or did not answer. The polls are at [http://www.spacecoalition.com]. A CNN/USA

Today/Gallup poll conducted at the same time as the 2005 poll for the Coalition found, however, that 40% of those surveyed wanted to set money aside for a human trip to Mars, while 58% were opposed [http://poll.gallup.com/content/default.aspx?ci=17224 and pg=1].

Congressional Reaction

NASA's oversight and funding committees (House Science, Senate Commerce, and House and Senate Appropriations[5]) have held many hearings on the Vision and related issues. Members of both parties have expressed support for the goal, but also concern about the cost and the impact on other NASA activities and NASA's workforce.

FY2005 Funding

For FY2005, Congress appropriated almost all the funding NASA requested for all its activities (see CRS Report RL32676 for more information on NASA's FY2005 budget). Many Vision supporters considered that action as a sign of congressional support. Conferees on the act, however, explicitly stated that although they were appropriating substantial funds for the Vision, "to date there has been no substantive Congressional action endorsing this initiative." (H.Rept. 108-792, p. 1599).

FY2006 Funding

NASA's total FY2006 budget request was $16.5 billion, a 2.4% increase over what was appropriated in the FY2005 Consolidated Appropriations Act (P.L. 108-447). If a $126 million FY2005 supplemental is added, NASA's FY2005 total was $16.2 billion, and the FY2006 request was a 1.6% increase. By comparison, the White House projected a 4.7% increase for FY2006 when it announced the Vision in 2004. CRS Report RL32988 describes NASA's FY2006 budget in more detail.

Within NASA's total budget request, what constitutes funding for the "Vision" is open to interpretation. A NASA budget chart from February 2005 divided the FY2006 request into three categories: "exploration-specific," $6 billion; "shuttle and space station," $6.4 billion; and "earth science, aeronautics, and others," $4.1 billion. Some may consider the Vision funding as that in the "exploration-specific" category, which combines funding for the Exploration Systems Mission Directorate (ESMD) with many programs in the Science Mission Directorate, such as robotic solar system exploration and space telescopes. Others may add funding for the space shuttle and space station, since those are often described as the first steps in the Vision. Alternatively, funding only for ESMD ($3.2 billion) could be defined as funding for the Vision, which includes development of the new Crew Exploration Vehicle and a launch vehicle for it.

Congress appropriated $16,456.8 million for NASA in the FY2006 Science, State, Justice Commerce (SSJC) appropriations act (H.R. 2862, P.L. 109-108). It provides a net increase of $500,000 above the requested level, but is subject to a 0.28% across-the-board rescission in that act, and a 1% rescission in another appropriations act (P.L. 109-148). In the SSJC act, Congress added approximately $500 million for specific projects or activities and cut a similar amount. For

ESMD, Congress cut $25 million from each of the two research and technology programs as recommended by the House; $34 million from Centennial Challenges (which awards monetary prizes for developing solutions to technological challenges set by NASA) as recommended by the Senate; and $200 million from Project Prometheus (see below). Congress also passed a NASA authorization act in 2005 (P.L.109-155). It endorsed the President's Vision, and authorized a total for NASA of $17.9 billion in FY2007 and $18.7 billion in FY2008. Both figures are higher than the out-year estimates in NASA's FY2006 budget request ($17 billion for FY2007 and $17.3 billion for FY2008). Congress continues to stress that NASA should continue a balanced set of programs, including science and aeronautics, not focus entirely on the Vision.

Comparison with the 1989 Space Exploration Initiative

Although some media stories portray President Bush's speech as the first new vision for NASA since the Apollo era, [6]President George H.W. Bush made a similar proposal on July 20, 1989, the 20th anniversary of the first human landing the Moon. Known as the Space Exploration Initiative (SEI), the senior President Bush directed that the space station serve as a stepping stone to returning humans to the Moon and someday sending them to Mars. In a May 1990 speech, he added more details, saying that he believed humans would reach Mars by 2019. Richard Darman, then Director of the Office of Management and Budget (OMB), stated at a press conference the day of the President's original speech that fulfilling the Moon/Mars goals would cost $400 billion over 30 years. Other estimates (some higher, some lower) were offered later by NASA.

SEI was announced during a period when Congress was attempting to reduce the federal deficit, and the proposal was not received enthusiastically. Funding for SEI was requested in the FY1991, FY1992, and FY1993 budgets, though what constituted "SEI funding" changed during those years. For FY1991 NASA requested $953 million for SEI. The FY1991 NASA authorization bill (P.L. 101-611) approved almost full funding, but the appropriations bill essentially zeroed it (P.L. 101-507), although NASA was allowed to reprogram $37 million into SEI for FY1991. For FY1992, the NASA request was $94 million; Congress approved $32 million. For FY1993, $64 million was requested. The FY1993 NASA authorization bill (P.L. 102-588) approved approximately half of that; the appropriations bill (P.L. 102-389) essentially zeroed it.

IMPLEMENTING THE VISION

Dr. Michael Griffin became NASA's Administrator in April 2005. In the early 1990s, while the senior President Bush was in office, Dr. Griffin headed NASA's SEI program (see above). Dr. Griffin's plans for implementing the Vision are somewhat different from those of his predecessor, Mr. Sean O'Keefe.

Crew Exploration Vehicle (CEV) and Crew Launch Vehicle (CLV)

President Bush directed NASA to develop a CEV to take astronauts to and from the Moon. He called for a demonstration flight by 2008, and an operational capability to Earth orbit no later than 2014. During Mr. O'Keefe's tenure, NASA would not commit to using the CEV to take crews to and from the ISS, wanting to keep the CEV design clearly focused on lunar transportation. Dr. Griffin, by contrast, has explicitly linked the CEV and the ISS. It will take crews back and forth to the ISS, and serve as a "crew return" vehicle (or lifeboat) for them. Mr. O'Keefe's plan was to open a competition for industry teams to design the CEV and conduct demonstration flights in 2008, after which a single contractor would be selected to build an operational CEV by 2014. Dr. Griffin decided that NASA itself should determine the basic design, and plans to award a single contract in 2006. He hopes for an Earth-orbital operational capability by 2012. That would leave a multi-year gap between the end of the shuttle and the availability of the CEV, during which time NASA would be dependent on Russia for transporting NASA astronauts to the ISS.

The CEV needs a launch vehicle, which is designated the Crew Launch Vehicle (CLV). The directive that NASA terminate the shuttle in 2010 refers to the shuttle in its current configuration — an airplane-like Orbiter carrying crew and cargo, with two Solid Rocket Boosters (SRBs) on each side, and a large, cylindrical External Tank (ET) containing fuel for the Orbiter's main engines. Dr. Griffin plans to develop two "shuttle-derived" launch vehicles (SDLVs) using the SRBs and ET, but not the Orbiter. One is the CLV, which will use one SRB and a new upper stage, with the CEV spacecraft on top. The other is a "heavy lift" SDLV using the ET and SRBs, with a cargo carrying spacecraft on top. For more information, see CRS Issue Brief IB93062.

Project Prometheus

NASA initiated this program to develop advanced nuclear power and propulsion systems before the Vision was announced. Its original goal was developing Nuclear Electric Propulsion (NEP) and advanced Radioisotope Thermoelectric Generators (RTGs) for a robotic probe, called the Jupiter Icy Moons Orbiter (JIMO), to study three of Jupiter's moons. RTGs have been used by NASA since the 1960s to supply power for spacecraft systems and experiments on probes that travel so far from the Sun that solar energy-based systems are impractical. RTGs also can be used for spacecraft that land on surfaces where they will experience "night" for long periods. NASA has not used nuclear propulsion, either NEP or a different type, Nuclear Thermal Propulsion (NTP), in the past, although NASA worked on developing NTP in the 1960s and early 1970s. After the Vision was announced, Prometheus took on greater importance because of its potential for providing power and propulsion for human exploration, as well as robotic spacecraft. However, the desire to accelerate development of the CEV means that funding must be shifted to it from other NASA activities. Project Prometheus is one of those. Initially, Dr. Griffin cut funds and refocused it on nuclear surface power systems (for use on the lunar surface, e.g.) and NTP, with NEP as a third priority. In a September 30, 2005 update to its operating plan, NASA announced it was further restructuring the program, yielding $76 million in FY2006 that would be redirected to the CEV/CLV program. JIMO was canceled, and only low-level funding will be provided for key nuclear systems research and technology issues. Funding for Prometheus was shifted back to the Science Mission Directorate (where it originated), although the program will be managed by ESMD. As noted, Congress cut $200 million from Prometheus in NASA's FY2006 appropriations act.

KEY ISSUES FOR CONGRESS

Many media accounts of the Vision focus on the long term "Moon/Mars" goals, but nearer term questions of how long to fly the space shuttle and utilize the International Space Station, and the funding impact on other NASA activities, are an immediate focus.

What Are the Implications of Terminating the Shuttle Program in 2010?

The shuttle program, and debate about its future, are discussed in CRS Report RS21408. The shuttle is the only U.S. vehicle capable of taking astronauts to and from space. President Bush directed NASA to retire the shuttle in 2010, when construction of the International Space Station (ISS) is expected to be finished. Dr. Griffin has repeatedly expressed his intention to terminate the shuttle program in 2010, whether or not ISS is completed. Issues include whether announcing a strict date for terminating the shuttle could create schedule pressure similar to that cited by investigators as a factor in the 2003 space shuttle *Columbia* tragedy; and whether the United States wants to be dependent on Russia for human access to space during the gap between the end of the shuttle program and the availability of the new CEV (see above).

Should U.S. Involvement in ISS End by 2016?

The ISS is discussed in more detail in CRS Issue Brief IB93017. It is being assembled in Earth orbit, and assembly is expected to be completed in 2010. Plans had called for ISS to be operated for at least 10 years after "assembly complete" as a laboratory for microgravity experiments, with research benefitting both future NASA exploration missions and people on Earth. Now, the plan is for NASA to complete its utilization of the ISS by 2016, and narrow its research program to only that needed to support human space exploration. Some question whether the U.S. investment in the space station (about $35 billion through FY2005) is worth the benefits under those circumstances. The NASA authorization act (P.L. 109-155) seeks to enhance the use of the ISS for scientific research by designating it as a national laboratory and requiring that 15% of ISS research funds be spent on non-Vision related research. Other questions include what will happen to the ISS when NASA ends its participation — will it be turned over to the other ISS partners, privatized, or deorbited?

What Are the Costs and Other Details?

The Bush Administration has not provided a cost estimate for the Vision as a whole, but in 2005, NASA estimated the cost for returning humans to the Moon by 2018 (its current goal) at $104 billion (excluding $20 billion that would be spent on ISS servicing using the CEV). A September 2004 Congressional Budget Office [http://www.cbo.gov] report cautioned that, based on historical trends at NASA, the actual cost could be much higher than NASA estimates.

Taking most of the requisite funds from other NASA programs instead of adding new money for the agency may mitigate concerns that the Vision could increase the deficit or detract from other national priorities. It raises issues, however, about the impact to those other NASA programs, and whether the level of funding is adequate to achieve the goals. Dr. Griffin asserts that he will not take money from space science, earth science, or aeronautics programs to pay for the Vision, so the source of the needed funds is unclear, especially since NASA's budget is projected to rise only at the rate of inflation for the foreseeable future. Additional questions include the extent to which other countries will want to participate in the Vision, and whether the Vision is an appropriate goal for NASA. Some argue that space exploration can be done more safely and less expensively by robotic probes.

REFERENCES

[1] A 1998 Intergovernmental Agreement (IGA), and Memoranda of Understanding between NASA and its counterparts, detail those obligations. See [[http://www.hq.nasa.gov/office/oer/]].

[2] The characterization of the $12.6 billion as an "addition" was based on a NASA assumption that without the Vision, NASA would have been held to a flat budget. The "increase," therefore, was above a hypothetical budget that is lower than what NASA projected in its FY2004 budget.

[3] The 2004 "sand chart" is at [http://www.nasa.gov/pdf/ 54873 main_budget_chart_14jan04.pdf].

[4] (1)"Results of AP Poll on Space Exploration," *Associated Press*, Jan. 2, 2004, 14:44. (2) Will Lester, "AP Poll: U.S. Tepid on Bush's Space Plans," *Associated Press*, Jan. 12, 2004, 14:50.

[5] For FY2005, NASA was overseen by the VA-HUD-IA subcommittees in the House and Senate appropriations committees. In the 109th Congress, NASA is part of the Science, State, Justice, Commerce (SSJC) appropriations subcommittee in the House; and the Commerce, Justice, Science (CJS) appropriations subcommittee in the Senate.

[6] The Apollo program was initiated by President John F. Kennedy in May 1961 to land a man on the Moon and return him safely to Earth before the end of that decade. NASA first developed experience with launching people

into space, and extravehicular activities (EVAs, or spacewalks), through the Mercury (1961-1963) and Gemini (1965-1966) programs. The first Apollo mission was to be launched in 1967, but the crew died on January 27, 1967 when a fire erupted in the Apollo command module during a pre-launch test. The first successful Apollo mission was launched in 1968, and the first Americans landed on the Moon on July 20, 1969 (Neil Armstrong and Buzz Aldrin, while pilot Michael Collins orbited above in the Apollo 11 spacecraft). A total of six two-man crews walked on the Moon from 1969-1972. Another crew (Apollo 13) intended to land on the Moon in 1970, but made an emergency return to Earth when the Service Module of their spacecraft exploded enroute to the Moon.

In: Space Policy and Exploration
Editor: W. N. Callmers, pp. 131-139
ISBN: 978-1-60456-448-8

Chapter 8

MILITARY SPACE PROGRAMS: ISSUES CONCERNING DOD'S SBIRS AND STSS PROGRAMS*

Marcia S. Smith

ABSTRACT

The Department of Defense's (DOD's) programs to develop new satellites to alert U.S. military commanders to foreign missile launches, and to support missile defense objectives, are controversial because of cost growth and schedule slippage. SBIRS-High, managed by the Air Force, would replace existing Defense Support Program "early warning" satellites. The Space Tracking and Surveillance System (STSS, formerly SBIRS-Low), managed by the Missile Defense Agency, would perform missile tracking and target discrimination for missile defense objectives. The SBIRS-High program has breached Nunn-McCurdy cost growth limits several times, resulting in several program restructurings.

* Excerpted from CRS Report RS21148, dated January 30, 2006.

SATELLITE EARLY WARNING SYSTEMS

The United States began developing early warning satellite systems in the 1950s to alert the National Command Authority to foreign missile launches. The current series is called the Defense Support Program (DSP). The first DSP satellite was launched in November 1970; 22 have been launched to date, most recently in February 2004. The final DSP was delivered to the Air Force in May 2005 for launch later in the year. Each DSP can operate for up to 10 years.[1] Four satellites reportedly are needed for a full operational capability. Six satellites reportedly were operating in January 2001.[2]

DSP satellites (built by Northrop Grumman Space Technology, which was formerly TRW Space and Electronics) use infrared sensors to detect the heat of fuel exhausts associated with missile launches. Sensors on the satellites also can detect nuclear bursts associated with the detonation of nuclear weapons. A February 2001 General Accounting Office (GAO, now the Government Accountability Office) report[3] recounted DOD's attempts to build a replacement for DSP over more than two decades. None of the proposed replacement programs — the Advanced Warning System in the early 1980s, the Boost Surveillance and Tracking System in the late 1980s, the Follow-On Early Warning System in the early 1990s, and the Alert, Locate and Report Missiles System in the mid-1990s — reached fruition "due to immature technology, high cost, and affordability issues," according to GAO. Instead, enhancements were made to the DSP series. For example, DSP was designed to detect launches of strategic long range missiles (such as Intercontinental Ballistic Missiles). However, the need to detect short range tactical missiles, such as Scud, was highlighted during the 1990-1991 Persian Gulf War. In 1995, DOD added the ALERT (Attack and Launch Early Reporting to Theater) system to DSP satellites to augment their theater missile warning capabilities.

DSP-type satellites are intrinsically part of any effort to develop a missile defense system because they provide the first warning that a foreign missile has been launched (during the missile's "boost" phase), but DSP also serves other objectives. Since the 1980s, there has been interest in developing a system explicitly to support missile defense — one that can track missiles as they progress along their flight path (the "mid-course" phase), detect and track warheads once they are deployed from the missile, and cue weapon systems to attack the missiles or warheads. A concept for a constellation of many satellites in low Earth orbit, called Brilliant Eyes, was developed during the 1980s under the auspices of the Strategic Defense Initiative Office (SDIO). Following a 1994 DOD study on how best to meet the nation's early warning needs, Brilliant Eyes was transferred to the Air Force, which was given responsibility to build an integrated Space-Based InfraRed System (SBIRS) with satellites in several orbits. Brilliant Eyes was renamed the Space and Missile Tracking System and became the low Earth orbit component of SBIRS. Later it was renamed SBIRS-Low. The system to replace DSP was named SBIRS-High, consisting of satellites in geostationary orbit (GEO, where DSP satellites are placed) and sensors on other DOD satellites in highly elliptical orbits (HEO).[4] In 2001, SBIRS-Low was shifted back to the Ballistic Missile Defense Organization (BMDO), which was SDIO's successor and is now the Missile Defense Agency. That action was taken to stress that SBIRS-Low's main purpose is to support ballistic missile defense. The satellite program has since been renamed again, and is now the Space Tracking and Surveillance System (STSS).

SBIRS-High

Purpose, Design, and Cost Estimate

SBIRS-High was designed to perform four missions: missile warning, missile defense, technical intelligence, and battlespace characterization (observing and reporting on military activities on a battlefield). The plan was for the system to consist of four operational GEO satellites (plus a ground spare), sensors on two classified DOD satellites in HEO, a ground-based Mission Control Station (MCS), and ground-based relay stations. MCS achieved initial operational capability in January 2002 using the existing DSP satellites.

In 1996, *Aviation Week and Space Technology* described the technical capabilities of SBIRS-High (November 18, 1996, p. 23) that were expected at that time. Reportedly it would have high speed scanning sensors and staring sensors. After the scanning sensor detected a launch, it would cue the staring sensor to observe the event and provide more detailed data. DSP satellites, by contrast, reportedly have only scanning sensors. DSP takes 40-50 seconds to detect a missile launch and determine its course, while SBIRS-High was being designed to make those determinations and relay warnings to ground forces in 10-20 seconds. The extent to which these technical design characteristics are still planned is unclear from the open literature.

A Lockheed Martin-Northrop Grumman team won a $2.16 billion contract to build SBIRS-High in 1996. In September 2002, DOD increased the contract to $4.18 billion, which does not include the cost of three of the five GEO satellites. DOD's Selected Acquisition Report for the quarter ending September 2005 [http://www.acq.osd.mil/ara/am/sar/2005-SEP-SST.pdf] shows the original cost estimate for SBIRS-High as $3.68 billion in FY1995 dollars ($4.15 billion in current year dollars, i.e., adjusted for inflation), compared with a new estimate of $9.01 billion in FY1995 dollars ($10.64 billion in current year dollars).

Issues

The SBIRS-High program is controversial because of cost growth and schedule slippage caused by technical challenges that have been encountered in developing the sensors and satellites. In the FY2002 DOD appropriations act, Congress denied all procurement funding ($94 million had been requested) because it felt more research and development (R and D) was required. It added $40 million to the $395 million requested for R and D. The House Appropriations Committee's report on the FY2002 DOD appropriations act (H.Rept. 107-298, p. 140) cited findings by GAO that the program was facing serious hardware and software design problems including sensor jitter, inadequate infrared sensitivity, and stray sunlight. (The GAO report is classified). *Space News* reported on January 7, 2002 (p. 14) that the program's cost estimate had grown from $1.9 billion to $4.5 billion, and the first launch slipped from 2002 to 2006. *Space News* attributed the cost increase to technical problems, including software development; faulty cost estimates; budget erosion; and schedule slippage.

In December 2001, DOD made the first of several notifications to Congress that SBIRS-High breached one of the two "Nunn-McCurdy" cost growth limits: a 15% Program Acquisition Unit Cost (PAUC) cost increase limit, which requires notification to Congress; or a 25% limit, where DOD must notify Congress and certify that the program is essential to national security, that no alternatives will provide equal or greater military capability at less cost, that new cost estimates are reasonable, and that the program management structure is adequate. The first breach violated the 25% limit, and DOD issued the required certification on May 2, 2002. The Air Force restructured the program. DOD decided to pay for the first two GEO satellites (then scheduled for launch in FY2006 and FY2007) using research, development, test, and evaluation (RDT and E) funds. The remaining three satellites would be purchased with procurement funds, with a two-year gap between the first two and the last three.

A May 2003 report of the Defense Science Board and Air Force Scientific Advisory Board [http://www.acq.osd.mil/dsb/reports/space.pdf] criticized early program management of SBIRS-High, and took a cautious attitude concerning whether the restructured program would succeed. An October 2003 GAO report

(GAO-04-48) concluded the program remained at "substantial risk of cost and schedule increases." The program experienced further setbacks. Delivery of the first SBIRS-High sensor, for launch on one of the HEO satellites, slipped to August 2004 because of electromagnetic interference between the sensor and other spacecraft equipment. (The second HEO sensor was delivered in September 2005.) Launch of the first GEO satellite slipped by two years; that launch is now expected in 2008. DOD notified Congress of additional Nunn-McCurdy breaches in June 2004 (of the 15% limit) and March 2005 (15% limit), and in July 2005 informed Congress than an independent program assessment concluded that the program exceeded the 25% limit (*Inside the Air Force*, August 5, 2005). DOD's Selected Acquisition Report for the period ending September 2005 indicates that the program exceeded the 15% Nunn-McCurdy limit during that reporting period. As noted above, the current cost estimate for the program is $10.64 billion in current year dollars. According to media accounts (e.g. *Aerospace Daily*, December 16, 2005, p. 1), DOD sent the required Nunn-McCurdy certification to Congress on December 12, 2005. While DOD still plans to buy the first two GEO satellites, it will buy, at most, one more. The funds that would have been spent for the fourth and fifth GEO satellites reportedly will be used instead to design an alternative system using state-of-the-art technologies.

FY2006 Budget

DOD requested $756 million. The FY2006 DOD authorization bill (P.L. 109-163) provides full funding. The FY2006 DOD appropriations act (P.L. 109-148) cut $50 million due to excessive program growth (the Senate had approved a $100 million cut). The House Appropriations report calls the program "extremely troubled" (H.Rept. 109-119, p. 181).

Space Tracking and Surveillance System (formerly SBIRS-Low)

Purpose, Design, and Cost Estimates

The Space Tracking and Surveillance System (STSS, previously SBIRS-Low) is designed to support missile defense. Management of the program was transferred from the Air Force back to the Ballistic Missile Defense Organization (BMDO, the successor to SDIO — see earlier discussion of Brilliant Eyes), to emphasize that missile defense is its primary objective. BMDO is now the Missile Defense Agency (MDA). For more on missile defense, see CRS Report RL31111, *Missile Defense: The Current Debate*, by Steven A. Hildreth.

The missile defense system is envisioned as a "layered" defense that can attack missiles or warheads in three different phases of flight: boost (launch), mid-course (enroute to a target, when warheads are deployed from the missile), and terminal (after reentry). The goal of an operational STSS is to track missiles through all three phases; discriminate between warheads and decoys; transmit data to other systems that will be used to cue radars and provide intercept handovers; and provide data for intercept hit/kill assessments. Tracking missiles during the mid-course phase is more difficult than during boost, because the missile is no longer firing its engines and hence does not have a strong infrared (heat) signature, making it necessary to track a cold object against the cold background of space. Similarly, tracking warheads after they have been deployed, and discriminating between warheads and decoys, is a technically challenging task.

Cost estimates are problematic because there is no final system architecture and the schedule is in flux. In its February 2001 report, GAO reported that DOD had estimated the life-cycle cost for STSS (then SBIRS-Low) through FY2022 at $11.8 billion. The House Appropriations Committee reported in late 2001 (H.Rept. 107-298, p. 250) that the program's life cycle cost had grown from $10 billion to over $23 billion. In March 2005, GAO reported (GAO-05-243) that DOD's estimate for the program between 2002 and 2011 is approximately $4.5 billion; a life cycle cost was not provided.

Two industry teams were chosen in 1999 for program definition and risk reduction (PDRR): Spectrum Astro/ Northrop Grumman, and TRW/ Raytheon. DOD was expected to select one of the teams for the next phase in mid-2002 and the satellites were to have been launched between 2006 and 2010.[5] In the April 2002 restructuring (see below), DOD merged the teams. Northrop Grumman Space Technology (formerly TRW Space and Electronics) is the prime contractor, and Spectrum Astro (later acquired by General Dynamics) is a major subcontractor, for building the satellites.

Issues

This program has gone through several name changes, making it difficult to track. Congress began expressing concern about it in 1996, when it was known as the Space and Missile Tracking System, particularly in terms of program management.[6] Indications of technical and funding problems publicly emerged in 1999 when DOD cancelled contracts with TRW and Boeing to build and launch three prototype demonstration satellites because of significant cost growth.[7] In the early 2000s, when its name was SBIRS-Low, questions arose as to whether it was vital to a missile defense system. Views vary. Some assert that missile defense cannot be achieved without such a system, while others argue that there are alternatives, such as ground-based radars.

To some extent, the answer may depend on the nature of the threat the missile defense system is expected to defeat (e.g., number of incoming warheads, or sophistication of countermeasures). Radars have been used for early warning of missile launches for decades, and already are envisioned as part of the missile defense system. To provide effective coverage, the radars might have to be based not only in the United States, but in other countries — radars in England and Greenland are part of the early warning system on which the United States relies today. The question is whether ground-based or sea-based radars can substitute for a space-based system, especially now that the United States is no longer bound by numerical and geographic limitations imposed on radars that are part of an anti-ballistic missile (ABM) system by the 1972 ABM treaty.

In its November 2001 report on the FY2002 DOD appropriations bill (H.R. 3338, H.Rept. 107-298, p. 250), the House Appropriations Committee cited an internal DOD study that indicated ground-based radars are a viable, lower cost, and lower risk, alternative.

Technical challenges continued. In its February 2001 report (cited earlier), GAO found that five of six critical satellite technologies were too immature to ensure they would be ready when needed. The House Appropriations Committee, in its November 2001 report (cited above), expressed concern that the program's life cycle cost had grown from $10 billion to over $23 billion. Consequently, the committee zeroed funding ($385 million was requested) and instead created a Satellite Sensor Technology program ($250 million) and a Ground Sensor Technology program ($75 million) as an alternative. Conferees approved the $250 million for Satellite Sensor Technology, but allowed the Secretary of Defense to spend it either on SBIRS-Low or new technology. It was spent on SBIRS-Low.

A restructuring plan was submitted to Congress on April 15, 2002. A system consisting of 20-30 satellites had been envisioned, with the first launch in 2006. The restructuring plan called for completing two "legacy" demonstration satellites that had been partially built as part of the 1999 plan, and launching them in 2006 and 2007. New technologies would be introduced in future satellites, and two new demonstration satellites would be launched beginning in 2010. In August 2002, DOD awarded Northrop Grumman Space Technologies an $869 million contract to complete the two legacy satellites, develop a ground system, and conduct preliminary engineering analysis of the new demonstration satellites, with options for building eight operational satellites. The program's name was changed to Space Tracking and Surveillance System (STSS). MDA modified the plan in late 2002 after it reduced STSS funding; GAO criticized many of those changes in a May 2003 report (GAO-03-597).

STSS is proceeding in a series of biennial "blocks." According to MDA's FY2006 budget documents, STSS' Block 2006 is the launch of the two legacy satellites; Block 2008 is an improvement of the ground system; and, in Block 2012, operational satellites would be integrated into the program. Block 2010, which appeared in FY2005 budget documents (without descriptive detail), is shown at zero in MDA's FY2006 documents. A March 2005 GAO report (GAO-05-243) states that the program content of Block 2010 and beyond is classified (and, thus, funding for Block 2010 could be in the classified budget). GAO reported that its review of STSS activities in 2004 did not find any indication that the two legacy satellites would not be able to launch on schedule.

FY2006 Budget

According to DOD's FY2006 budget documentation, $232 million was requested for STSS in the unclassified budget. No changes were specified in the DOD authorization bill (P.L. 109-163). In their reports on the FY2006 DOD appropriations bill (H.R. 2863, H.Rept. 109-119, S.Rept. 109-141), the House and Senate Appropriations Committees restructured MDA's budget, moving this program out of the BMDS Sensors line and creating a new Space Surveillance and Tracking System (SSTS, instead of STSS) line. The committee reports show the requested funding level as $235.5 million. The final version of the bill (P.L. 109-148) cuts $538,000 from SSTS, the entire amount requested for Block 12, stating it is a "premature request." It also adds $5 million for the airborne infrared surveillance system (AIRS) under this line item.

References

[1] *Space News*, Jan. 7, 2002, p. 14.

[2] *Space and Missile Defense Report*, Jan. 18, 2001, p. 8. More recent data could not be obtained in the open literature.

[3] U.S. General Accounting Office. Defense Acquisitions: Space-Based Infrared System-low at Risk of Missing Initial Deployment Date. GAO-01-

6. Washington, U.S. GAO, Feb. 2001. For more detail on the history of U.S. early warning satellite systems, see Richelson, Jeffrey. "America's Space Sentinels." Lawrence, Kansas, University Press of Kansas, 1999.

[4] Geostationary orbit (GEO) exists 35,800 kilometers above the equator. A satellite in GEO maintains a fixed position relative to a point on Earth. Three or four properly spaced GEO satellites can view the entire globe, except for the polar regions. HEO orbits can provide coverage of the polar regions. A classic HEO orbit (called a Molniya orbit after the Soviet communications satellite system that first utilized it), has an apogee (the highest point of the orbit) of approximately 40,000 kilometers, and a perigee (the lowest point) of about 500 kilometers, giving the orbit an elliptical shape. With an inclination of about 63 degrees (the angle at which it intersects the equator), such an orbit allows a satellite to linger or "dwell" over the northern hemisphere for several hours per orbit, viewing parts of the globe not observable from GEO. DOD reportedly uses this type of orbit for classified satellites.

[5] The first launch was scheduled for 2006, but Congress directed DOD to accelerate the schedule to 2002. The Defense Science Board concluded that 2002 was technically feasible, but 2004 would represent a more efficient approach. (*Aerospace Daily*, Oct. 3, 1996, p. 18; *Space News*, Sept. 16-22, 1996, p. 10.) The date then slipped back to 2006 primarily due to funding issues.

[6] Authorizers Blast DOD for SMTS Management. *Aerospace Daily*, Aug. 8, 1996, p. 207.

[7] Ferster, Warren. SBIRS Demonstration Projects Terminated. *Space News*, Feb. 15, 1999, p. 1.

In: Space Policy and Exploration
Editor: W. N. Callmers, pp. 141-148
ISBN: 978-1-60456-448-8

Chapter 9

HUBBLE SPACE TELESCOPE: SHOULD NASA PROCEED WITH A SERVICING MISSION?*

Daniel Morgan

ABSTRACT

The National Aeronautics and Space Administration (NASA) estimates that without a servicing mission to replace key components, the Hubble Space Telescope will cease scientific operations in 2008 instead of 2010. In January 2004, then-NASA Administrator Sean O'Keefe announced that the space shuttle would no longer be used to service Hubble. He indicated that this decision was based primarily on safety concerns in the wake of the space shuttle *Columbia* accident in 2003. Many critics, however, saw it as the result of the new Vision for Space Exploration, announced by President Bush in January 2004, which focuses NASA's priorities on human and robotic exploration of the solar system. Hubble supporters hope to reverse the decision and proceed with a shuttle servicing mission. Dr. Michael Griffin, who became NASA Administrator in April 2005, has stated that he will reassess whether to use the shuttle to service Hubble after there have been two successful post-*Columbia* shuttle flights. Problems during the first flight in July 2005 led to postponement of the second, which is now planned for July 2006.

INTRODUCTION

NASA launched the Hubble Space Telescope in 1990 aboard the space shuttle *Discovery*. Unlike other NASA space telescopes, Hubble was designed to be serviced regularly by astronauts. That design proved fortuitous when it was discovered that Hubble had a defective mirror that produced blurry images. Astronauts on the first servicing mission in 1993 were able to install corrective optics, allowing years of scientific accomplishments and generating widespread scientific and public support. Additional servicing missions were conducted in 1997, 1999, and 2002 to replace aging hardware and install advanced scientific instruments. Two more shuttle missions to Hubble were scheduled: another servicing mission in 2004 (known as SM-4) and a retrieval mission to bring the telescope back to Earth in 2010. Following the space shuttle *Columbia* accident in February 2003, however, then-NASA Administrator Sean O'Keefe decided not to proceed with either flight. Current NASA Administrator Michael Griffin plans to revisit that decision once the shuttle returns to regular flight.

THE HUBBLE SPACE TELESCOPE

Roughly the size of a school bus, the Hubble Space Telescope was designed to make astronomical observations of the universe in the visible, ultraviolet, and near-infrared wavelength bands. Although ground-based telescopes can also make visible and infrared observations, Hubble's location above the Earth's atmosphere enhances image clarity, enabling astronomers to look at fainter, more distant objects and further back in time. Hubble is operated for NASA by the Space Telescope Science Institute (STStcI) near Baltimore, MD. Websites maintained by STScI ([http://hubblesite.org/]) and NASA ([http://hubble.nasa.gov]) provide information about the telescope and its discoveries.

During servicing visits to Hubble, shuttle crews repair or replace aging equipment and install updated scientific instruments. Hubble has six gyroscopes for pointing the telescope, but two are now nonfunctional and one has degraded performance. Until recently, three were required to achieve the accuracy needed for scientific observations. New techniques implemented in August 2005 allow operation on just two gyroscopes, so that one is kept in reserve. Solar arrays generate electricity for the telescope; the energy is stored in batteries. Hubble has no propulsion system, relying instead on the space shuttle to boost its orbit so that

* Excerpted from CRS Report RS21767, dated June 29, 2006.

it does not reenter Earth's atmosphere. The tasks for the SM-4 mission included replacing all the gyroscopes and batteries and a fine guidance sensor, emplacing new thermal protection blankets, boosting Hubble's orbit, and installing two new scientific instruments (the Cosmic Origins Spectrograph for ultraviolet observations of chemical composition, and the Wide Field Camera 3 for observations from ultraviolet through near-infrared).

Hubble was designed to operate for 15 years, a milestone that was reached on April 25, 2005. NASA had planned to extend the operational period until 2010, at which time the space shuttle would bring Hubble back to Earth to prevent an uncontrolled reentry that could pose a debris risk to populated areas. Funding constraints, however, led some NASA officials to conclude that Hubble's operations should be brought to end earlier than 2010. The "funding wedge" created by ending the servicing missions would be used to build the new James Webb Space Telescope (JWST), which is being designed for infrared observations. The debate over how long to continue to operate Hubble, and linkage with funding for the JWST, was under way at the time of the *Columbia* accident.

The *Columbia* Accident and the Decision to Cancel SM-4

On February 1, 2003, the space shuttle *Columbia* disintegrated as it returned to Earth following a 16-day scientific mission. All seven astronauts aboard perished. The shuttle system was immediately grounded. NASA established the *Columbia* Accident Investigation Board (CAIB) to determine the causes of the accident and recommend corrective actions. (CRS Report RS21408, *NASA's Space Shuttle Program: The Columbia Tragedy, the Discovery Mission, and the Future of the Shuttle*, by Marcia S. Smith, discusses the *Columbia* accident.)

It was quickly apparent that SM-4 would be delayed. As CAIB deliberated, NASA decided that the 2010 Hubble retrieval mission was too risky compared to the benefits.

The agency began investigating alternatives for a controlled deorbit of the telescope so that any debris would fall in an unpopulated area such as the Pacific Ocean.

On January 16, 2004, Mr. O'Keefe informed workers at STScI and NASA's Goddard Space Flight Center (which built Hubble and oversees STScI) that he was canceling SM-4. According to the director of STScI, Dr. Steven Beckwith, Mr. O'Keefe cited several factors: the shuttle would not have the ISS as a safe haven; the changes required to meet the CAIB's recommendations regarding non-ISS related shuttle missions would not have application beyond the servicing mission, making their expense questionable; completing ISS construction by 2010 will require all the shuttle flights in that time period; Hubble's life would be extended for only a few years; and astronomers have other ground-and space-based telescopes they could use.[1]

Reaction to the Decision

Two days before Mr. O'Keefe announced his decision, President Bush directed NASA to embark on a new exploration initiative, requiring a shift in program and funding priorities. (See CRS Report RS21720, *Space Exploration: Issues Concerning the "Vision for Space Exploration"*, by Marcia S. Smith.) Funding for the new initiative would come primarily from canceling, deferring, or delaying other NASA programs. Although Mr. O'Keefe stated that the Hubble decision was based primarily on shuttle safety concerns, the timing of his announcement led many commentators to conclude that it was linked to the priority shifts required by the President's initiative. While some media accounts praised the NASA Administrator for making a difficult decision, others called Hubble the first "victim" of the President's initiative and chided NASA for putting the new exploration goals ahead of the astronomical research performed with Hubble.[2]

Servicing Options

Initial opposition to the cancellation of SM-4 focused on attempts to reverse Mr. O'Keefe's decision and proceed with a shuttle mission. The debate centered on comparing the risk of a mission to Hubble with the risk of a mission to the ISS. Shortly after the cancellation decision, NASA's then-Chief Scientist, Dr. John Grunsfeld, an astronaut who was a member of the 1999 and 2002 Hubble servicing crews, commented that if a shuttle mission to Hubble were mounted, it would be necessary to have a second shuttle ready to launch in case the first one encountered difficulties.[3] NASA had a backup shuttle ready for the first shuttle launch after return to flight, and plans to have one for the second, but not for subsequent missions if safety modifications appear to be working well.

Attention soon shifted to robotic servicing options, which dominated the public discussion of Hubble's future throughout most of 2004. At a Senate

Appropriations VA-HUD-IA Subcommittee hearing on March 11, 2004, Mr. O'Keefe agreed with a request from Senator Mikulski to ask the National Research Council (NRC) to study options for extending Hubble's life, including both shuttle and robotic missions.

In November 2004, Congress passed the FY2005 Consolidated Appropriations Act (P.L. 108-447), which provided $291 million for a Hubble servicing mission. The conference report (H.Rept. 108-792) stated that "a successful servicing mission to Hubble should be one of NASA's highest priorities." The report language did not specify whether the servicing mission should involve the space shuttle or a robotic mission.

The final NRC report on servicing options, released on December 8, 2004, found it "unlikely that NASA will be able to extend the science life of [Hubble] through robotic servicing" and that the risk of a shuttle mission to Hubble is similar to the risk of a single shuttle mission to the ISS. The report recommended a shuttle servicing mission, and a robotic mission only for deorbiting the telescope at the end of its useful lifetime.[4] Robotic servicing now appears to be off the table.

Shuttle Servicing Revisited

Dr. Michael Griffin was sworn in as NASA Administrator, replacing Mr. O'Keefe, on April 14, 2005. At his Senate confirmation hearing on April 12, 2005, Dr. Griffin stated that he was very familiar with the robotic servicing option and did not think it was feasible within the required time frame. He added that he will revisit the question of whether to use the shuttle to service Hubble after the second successful post-RTF shuttle flight, at which time NASA will be able to assess the risk factors associated with "essentially a new vehicle". The NASA Authorization Act of 2005 (P.L. 109-155), enacted in December 2005, calls for a shuttle servicing mission after the shuttle returns to flight successfully "unless such a mission would compromise astronaut safety" and requires a status report on servicing plans within 60 days of the landing of the second successful post-RTF flight. Problems during the first post-RTF flight in July 2005 led to postponement of the second, which is now planned for July 2006.

Because Dr. Griffin's decision to reconsider a shuttle mission to Hubble came after the release of the FY2006 budget, its cost was not included in the budget request for that year. A budget amendment submitted on July 15, 2005, requested an additional $30 million for Hubble, but this was only to preserve the option of a shuttle servicing mission, and was not intended to reflect the full cost of servicing

Hubble if a decision is made to proceed. In November 2005, Congress appropriated a total of $271 million for Hubble: the original request, plus the addition requested in the July budget amendment, plus another $50 million "to continue planning, preparation, and engineering activities" for SM-4, "pending a final decision" to proceed (H.Rept. 109-272, conference report for P.L. 109-108). The 2005 authorization act (P.L. 109-155) did not specify Hubble funding.

The FY2007 budget request states that it includes full funding for a shuttle-based servicing mission to Hubble in early FY2008, but it does not specify a dollar figure. The total FY2007 request for Hubble, including ongoing operations as well as preparations for a servicing mission, is $336.7 million. According to the Government Accountability Office (GAO), NASA estimated in late 2004 that a shuttle servicing mission to Hubble, including the full cost of a shuttle launch, would cost between $1.7 billion and $2.4 billion. GAO described this estimate as "not yet definitive", however.[5] It is considerably more than pre-*Columbia* estimates that were in the vicinity of $100 million plus the marginal cost of a shuttle launch.

Deorbiting Hubble

Plans for deorbiting Hubble also remain uncertain. Before the *Columbia* accident, the space shuttle was to return Hubble to Earth at the end of its lifetime, but there is no longer any expectation of retrieval by the shuttle, even if there is a shuttle servicing mission. Hubble has no propulsion system of its own, however, so a propulsion module must be attached to the spacecraft to permit a controlled deorbit. Remotely attaching the required new hardware is technologically challenging. If there is a shuttle servicing mission, a deorbit module could be attached at that time. However, analysis by NASA during 2005 indicates that Hubble is unlikely to make an uncontrolled reentry until at least 2020, rather than 2012 as previously believed, and the agency now considers the deorbiting issue to be "beyond the budget horizon."[6] Boosting Hubble's orbit during a shuttle servicing mission could delay the date of reentry even further, but at some point, deorbiting will be necessary if an uncontrolled reentry is to be avoided.

KEY QUESTIONS FOR CONGRESS

If there is no successful servicing mission, Hubble's lifetime is likely to be limited by the failure of its gyroscopes and batteries.[7] Both systems wear out over time, and historical experience permits projections of their future performance. The NRC study estimated that science operations would likely cease in about September 2007 as the result of gyroscope failures. Now that Hubble is operating on two gyroscopes rather than three, NASA believes that this date can be extended until mid-2008.[8] Servicing could still restart operations after science operations cease. However, the NRC also estimated that battery degradation would result in permanent structural damage in about May 2011, and perhaps as soon as December 2009 in the worst case. (A minimum level of battery power is required to keep Hubble warm and prevent warping of the optical system.) Servicing after this date would not be successful. NASA estimates that if left unattended, Hubble will make an uncontrolled reentry in about 2020.

In June 2006, it seemed that Hubble science might be limited sooner than expected by problems with Hubble's main camera, the Advanced Camera for Surveys (ACS). The ACS was taken off line on June 19, 2006, to investigate problems with its power supply electronics. NASA now anticipates, however, that these problems will be fully resolved in early July 2006.[9]

As Congress continues to review the Hubble servicing issue, the following questions may arise:

- How much would a shuttle servicing mission cost? What other programs would be cut to pay for it? (For example, to offset the funding that Congress provided for Hubble in FY2005, NASA's May 2005 operating plan postponed two other astronomy missions and reduced funding for Mars exploration.) What is the risk to the shuttle astronauts? Is the science to be gained worth the cost and risk?
- If Hubble is serviced, it is expected to operate for about five more years, rather than being deorbited in 2010 as previously planned. How much will operations after 2010 cost? (The FY2007 budget request, which was the first to extend its out-year estimates to FY2011, estimated Hubble costs in that year of $138.5 million.) How might those costs affect funding for other astronomy missions, such as JWST?
- To what extent could astronomers use ground-based telescopes instead of Hubble? To what extent could they use other space telescopes, such as JWST or the Spitzer Space Telescope? To what extent could they rely on

analyzing the archive of data already collected by Hubble? (Approximately five years of Hubble data have yet to be analyzed.)

REFERENCES

[1] See [http://www.stsci.edu/resources/sm4cancellation/sm4meeting.html].

[2] For example, see "Sacrificing Hubble," *Space News*, Jan. 26, 2004, p. 12; and Brian Berger, "Canceled Hubble Repair the First Victim of New NASA Vision," *Space News*, Jan. 26, 2004, p. 6.

[3] Quoted in Brian Berger, "Canceled Hubble Repair the First Victim of New NASA Vision," *Space News*, Jan. 26, 2004, p. 6.

[4] National Research Council, *Assessment of Options for Extending the Life of the Hubble Space Telescope*, National Academies Press, 2005. Online at [http://books.nap.edu/catalog/11169.html].

[5] Government Accountability Office, Space Shuttle: Costs of Hubble Servicing Mission and Implementation of Safety Recommendations Not Yet Definitive, GAO-05-34, Nov. 2004.

[6] NASA FY2007 budget justification, p. SAE SMD 3-20.

[7] In some scenarios, the limiting factors might instead be the failure of Hubble's fine guidance sensors or its avionics system.

[8] "Hubble Space Telescope Begins 'Two-Gyro' Science Operations," NASA press release, August 31, 2005.

[9] "NASA Issues Hubble Space Telescope Status Report," NASA press release, June 27, 2006.

INDEX

A

B

C

D

E

F

G

H

I